Toxic Bodies

TOXIC BODIES

Hormone Disruptors and the Legacy of DES

Nancy Langston

Yale University Press / New Haven and London

Published with assistance from the
Mary Cady Tew Memorial Fund.

Set in Galliard type by Keystone Typesetting, Inc.,
Orwigsburg, Pennsylvania.
Printed in the United States of America.

The Library of Congress has cataloged the
hardcover edition as follows:
Langston, Nancy.
Toxic bodies : hormone disruptors and the legacy of DES /
Nancy Langston.
p. ; cm.
Includes bibliographical references and index.
ISBN 978-0-300-13607-4 (alk. paper)
1. Endocrine disrupting chemicals — History.
2. Endocrine disrupting chemicals — Government policy —
United States — History. I. Title.
[DNLM: 1. Endocrine Disruptors — adverse effects —
United States. 2. Endocrine Disruptors — history — United
States. 3. Environmental Exposure — adverse effects —
United States. 4. Environmental Exposure — history —
United States. 5. History, 20th Century — United States.
6. History, 21st Century — United States.
WK 11 AA1 L285t 2010]
RA1224.2.L36 2010 615'.36 — dc22 2009025238

ISBN 978-0-300-17137-2 (pbk.)

A catalogue record for this book is available from the
British Library.

10 9 8 7 6 5 4 3 2

CONTENTS

PREFACE

In 2000, I invited Maria, a graduate student at the University of Wisconsin–Madison, to address my undergraduate environmental studies seminar. Maria had grown up along the Fox River in Wisconsin, where paper mills lined the shore. During her childhood, the stench from the mill waste in the river had been so bad that the city of Green Bay had dumped perfume in the water. But perfume could not mask the toxic contamination. In the 1960s the paper companies had manufactured carbonless copy paper coated with industrial chemicals known as polychlorinated biphenyls (PCBs). Few scientists had suspected the potent hormonal effects that PCBs could have on developing fetuses and children, and the chemicals had gone essentially unregulated. Many of the PCBs used by the paper companies had made their way into the Fox River, where they had accumulated in the fatty tissues of fish.

Every Friday night, Maria's family had participated in the Wisconsin tradition of the fish fry, going to a tavern to eat their fill of local fish. On hot summer days they had splashed in the cool waters of Green Bay, where the Fox River empties into Lake Michigan. And now, decades later, the river was a Superfund site where various groups contested responsibility for cleaning up the PCBs, chemicals that had become notorious for their toxic properties, particularly their ability to disrupt hormone systems.

Although Maria was training as an environmental scientist, she did not talk to my class about the technical details of her research. She did not dwell on the hormonal effects of parts-per-billion of PCBs. She did not describe the ways that PCBs changed thyroid hormone function or the ways that the chemicals altered brain development and the immune system. Instead, she talked with us about her young child. Should she breastfeed her daughter, she asked us. Maria knew from her research that, as an infant, her child was particularly vulnerable to chemicals that had accumulated into more concentrated and toxic forms. Breastfeeding would reduce the concentration of PCBs in Maria's own body, accumulated over decades lived along the Fox River. But she would pass on those chemicals to her daughter, with unknown and potentially tragic effects. Knowing that her own body was a toxic waste site, how could she breastfeed her child? At the same time, knowing that breast milk offered many health benefits to babies, how could she deny those to her daughter?

Like the rest of my class, I was haunted by Maria's dilemma. The thought that we have saturated rivers, wildlife, and ourselves with synthetic chemicals with potentially toxic effects began to trouble me. Maria's story gave a human face to the accumulating data suggesting that reproductive problems are increasing across a broad range of animals, from Great Lakes fish to people. Many researchers suspect that the culprits are synthetic chemicals that disrupt hormonal signals, particularly in the developing fetus. In the past decade, thousands of experimental studies have shown that synthetic chemicals can alter hormones in laboratory animals and wildlife, while numerous human studies have found correlations between exposure to industrial chemicals and reproductive problems. Endocrine-disrupting chemicals are not rare; they include the most common synthetic chemicals in production. Since World War II, synthetic chemicals in plastics, pharmaceuticals, and pesticides have permeated bodies and ecosystems throughout the United States, often with profound health and ecological effects, yet the government has largely failed to regulate them. How has this massive regulatory failure come about? Given what has been known about the risks of endocrine-disrupting chemicals since the 1940s, why have federal regulatory agencies done so little to protect public and environmental health?

Industry advocates argue that government bureaucracies have held back progress by overregulating chemicals, banning or limiting their use without scientific proof of harm. Environmentalists counter that both laws and ethics forbid the human experiments that would provide that proof. In the absence of complete knowledge, environmentalists argue, what is known as the precautionary principle should guide the regulation of toxic chemicals: if an action might cause severe or irreversible harm to complex systems, the burden of proof should be on the industry to show that it is safe, rather than on affected communities to show that it is harmful.

What can history teach us about scientific uncertainty and the precautionary principle that can help guide the regulation of endocrine disruptors? In *Toxic Bodies,* I examine the histories of several key endocrine-disrupting chemicals, including the synthetic estrogen diethylstilbestrol (DES), various pesticides such as DDT (Dichloro-Diphenyl-Trichloroethane), and several compounds found in common plastics. In each of these cases, scientists had substantial cause for health concerns when the chemicals were introduced, yet in each case, federal agencies were slow to protect public health.

The most detailed case study in this book focuses on DES, the first synthetic chemical to be marketed as an estrogen and one of the first synthetic chemicals identified as an endocrine disruptor. Beginning in the 1940s, millions of women were prescribed DES by their doctors, at first to treat the symptoms of menopause. In 1947 the Food and Drug Administration (FDA) approved DES for pregnant women with diabetes, and drug companies advertised it widely, promoting the use of DES in *all* pregnancies as a way to reduce the risk of miscarriage. Although no evidence ever supported this claim, millions of pregnant woman took the drug.

Meanwhile, millions of Americans were also being exposed to DES through their diet. Beginning in 1947, DES was approved in the United States as a steroid to promote growth, first in poultry and then in cattle. High levels of DES were soon detected in poultry sold for human consumption—up to a hundred times the concentration necessary to cause breast cancer in mice. Concern over DES's effects soon grew among women who used the drug, farmers who handled treated livestock, and workers who manufactured the material. Federal agencies initially dismissed these concerns as unfounded, but eventually, after exposed male agricultural workers suffered sterility, impotence, and breast growth, the FDA banned the use of DES implants in chicken in 1959, while allowing its continued use in cattle feed and for pregnant women.

The chemical became an environmental issue as well as a personal health issue. By the 1950s, farmers gave cattle the hormone to promote rapid weight gain, which was a key factor enabling the rapid expansion of industrialized feedlots. As cattle excreted waste, the metabolic byproducts of DES moved from feedlots into broader ecosystems, exposing a wide range of wildlife to the hormone. Chemical residues in the food supply changed the internal ecosystems of humans, livestock, and wildlife, interconnecting their bodies with their environment in increasingly troubling ways.

In 1971 researchers in Boston reported a cluster of extremely rare vaginal cancers in young women whose mothers had taken DES while they were pregnant. These problems had not been apparent at birth; they emerged only at puberty or young adulthood, sometimes decades after fetal exposure. Mothers and children exposed to DES organized to call for research into the drug, and eventually consumers, scientists, and concerned congressional representatives forced the FDA to ban the chemical for most uses.

The full dimensions of the health and environmental disaster that resulted from widespread DES use are only now becoming apparent. By 2002, DES had emerged in toxicological studies as a carcinogen and developmental toxicant so potent that the toxicity of other chemicals is often measured against it. Of the two to five million children who were exposed to DES prenatally, nearly 95 percent of those sampled have experienced reproductive-tract problems, including menstrual irregularities, infertility, and higher risks of a variety of reproductive cancers. At the peak of its use in the 1960s, DES was given to nearly 95 percent of feedlot cattle in the United States, which meant that millions of people consumed meat tainted with the artificial estrogen, and the estrogenic wastes from feedlots made their way into aquatic ecosystems, with unknown effects.

Why did the FDA approve the drug? Even before the agency approved DES in 1941, researchers knew that it caused cancer and problems with sexual development in laboratory animals. These concerns initially led FDA commissioner Walter Campbell to reject the drug in 1940, arguing that regulators must follow what he called the "conservative principle," essentially adopting the precautionary principle sixty years before that term came into common usage. Yet a year later, the FDA abandoned its position of precaution, and by 1947 the agency was telling critics of DES that it was up to them to prove that DES had caused harm, rather than up to the drug companies to show that it was safe. When companies applied for approval to use DES in livestock and for pregnant women, the same pattern unfolded twice more. Each time, the agency refused approval, citing the need for precaution given the known risks of the drug. But each time, it quickly gave way to industry pressure.

Why were regulators unable or unwilling to resist industry pressure? To begin with, they tended to share certain cultural and conceptual beliefs that industry lobbyists were quick to exploit. Contemporary scientific models of toxicology and development generally did not allow for the possibility that very low levels of synthetic chemicals could influence hormonal actions in the body. Indeed, emerging research that showed the harmful effects of various synthetic chemicals often seemed to violate the standard toxicological paradigms of the era, making it difficult for regulators to interpret scientific results. Even when experimental evidence from laboratory animals seemed to provide compelling proof of harm, uncertainty about the validity of animal studies in assessing risks for people made it difficult for regulators to defend principles of precaution in court.

In addition, confusion about the boundaries between natural and synthetic chemicals made it difficult for agency staff to understand why synthetic chemicals might cause harm even though natural estrogenic chemicals, which were relatively common in the food supply, did not. Cultural assumptions about gender differences also shaped the ways that scientists, regulators, and consumers understood hormones and their effects on the body. Finally, many regulators shared with industry staff a modernist worldview that combined faith in scientific expertise with the belief that technological progress could and should control nature. These beliefs often made regulators more skeptical of consumer claims of harm than they were of industry claims of safety. And while individual staff members within the federal agencies worked hard to protect public health, political appointees who headed the agencies often seemed more responsive to industry concerns about profits than to their own staff's concerns about risks.

Political, cultural, and scientific pressures all shaped these repeated retreats from precaution, and the echoes of those decisions still haunt us today. Menopausal and pregnant women no longer take DES, and livestock are no longer fattened with it, but Americans still face problems posed by chemicals with similar hormonal effects. Livestock continue to be treated with steroids, while pesticides continue to proliferate in the food supply. Plastics such as bisphenol A leach chemicals with hormonal activity into the drinking water, and every month brings new reports of intersex fish and cancer-ridden whales. And the effects of DES exposure still confront many people, including the women who took the drug, their sons and daughters who were exposed to it in the womb, and the people who ate meat tainted with its residues.

As I worked on this book, I found myself exploring a parallel toxic history of my own possible exposure to DES. Like 20 to 80 percent of American women, I had uterine leiomyomas, tumors commonly known as fibroids. Estrogen stimulates these tumor cells, and fetal exposure to DES has been linked to fibroid growth. Midway through the research, my formerly benign fibroids suddenly became much more problematic. Doctors poked me up and down, debating whether the unusually rapid tumor growth meant I had uterine cancer rather than just fibroids. They speculated about whether my cervical dysplasia, deformed uterus, and tumors might be signs of DES exposure. After a hysterectomy, when the pathologist's report indicated that "no normal uterine tissue" had been found in my "grossly malformed" uterus, I realized that a book I had thought was about other people's experience might indirectly be about my own.

My typical bias is to be skeptical. I have been trained, first as a scientist and then as a historian, to distrust correlations as causal explanations and to be leery of those who see disaster everywhere. But my perspective shifted slightly as I began thinking about my own family's cancers and the toxic chemicals present in the places where I grew up on the East Coast and where I now make my home in Wisconsin.

I started wondering how many of my family's cancers and reproductive problems might be linked to toxic exposures. My mother had colon cancer; my grandmother died of melanoma; a great aunt had breast cancer; another aunt died of pancreatic cancer. Of the five women in my immediate family, all have had a string of reproductive issues: miscarriages, infertility, endless fibroids, diseased fallopian tubes, ectopic pregnancies, cervical dysplasia, hysterectomies, excised ovaries, suspicious mammograms and breast biopsies, and two cases of suspected uterine cancer. Each of these is linked, in laboratory studies on animals and in epidemiological studies on women, to endocrine disruptors. But not a single one of us can point to a specific reproductive problem and pinpoint a specific exposure as the cause. Any given reproductive failure could have been random bad luck. Any given exposure was probably harmless. Nevertheless, we have somehow created a world in which mushrooming chemical exposures go hand in hand with reproductive chaos, partly because no one can prove that an individual chemical caused a particular health problem, and so regulators have largely failed to act.

In this book I examine the landscape of exposure that begins in our own bodies and connects us across generations, across species, and across ecosystems. Failures of regulation are expressed not just in hearings and court cases, but also inside our own bodies. How did Americans persuade themselves after World War II that it was a good idea to release millions of tons of chemicals known to be toxic into the environment? What assumptions about scientific expertise, the role of experts, and vulnerability to natural fluctuations drove the postwar generation's faith in better living through chemistry? How did we come to accept the increasingly toxic saturation of our bodies and our environments? How have cultural constructions of sex and gender shaped scientific and policy responses to endocrine disruptors? Diethylstilbestrol has largely been banned, but thousands of other endocrine disruptors remain in common use. Learning the lessons of DES can help us address current disputes over regulating today's endocrine-disrupting chemicals.

ACKNOWLEDGMENTS

Maria Powell's generosity with her story and her willingness to take my class up to the Fox River motivated this book. I am grateful to her. I also owe a great debt to Virginia Scharff and Nancy Scott Jackson, who encouraged me to undertake this project.

Colleagues and students in numerous seminars gave me thoughtful criticism and feedback at many stages of this project. In particular, I would like to thank students and faculty at the Land Institute, the Hall Seminar on the Humanities at the University of Kansas, Brandeis University, the University of Wisconsin–Oshkosh, Lawrence University, the seminar on Environmental and Agricultural History at the Massachusetts Institute of Technology, UCLA, the University of California–Davis, Dartmouth College, the University of Iowa, the University of Wisconsin's Center for Culture, History, and Environment, and the Agrarian Studies Seminar at Yale University. All were generous with their thoughts and suggestions. Conversations with Bill Cronon, Brian Donahue, Michael Egan, Lynne Heasley, Gregg Mitman, Jody Roberts, Ellen Stroud, Sarah Vogel, and Louis Warren were important in helping me develop my ideas and frame my arguments.

Without the willing and tireless assistance of staff at various archives and records depositories, I could never have re-created the narratives of the FDA staff. Archivists at the National Archives and Records Administration at College Park, Maryland, and staff at the Freedom of Information Office of the Food and Drug Administration were especially helpful.

Conversations with scientists involved in endocrine-disruption research helped me understand emerging research. I would like to thank Tyrone Hayes, John McLachlan, and Pete Myers in particular. Sandra Steingraber was generous with her comments, and the sociologist Susan Bell kindly shared her ideas and papers with me early in this project.

A sabbatical leave from the University of Wisconsin–Madison helped me begin research on this book, and time spent at the Christine Center enabled me to write about difficult issues. A fellowship from the American Council for Learned Societies helped fund the research, and summer research grants from the University of Wisconsin's graduate school supported the writing. I would especially like to thank my friends along Lake Superior who have made the north woods such a wonderful place to spend summers writing.

I own an enormous debt to my mother, Joann Langston, who read a draft of the manuscript with great care. Throughout the research process, she always gave me a place to stay near the National Archives, and she shared her decades of experience in federal agencies with me. Growing up in a suburb of Rockville, Maryland, where nearly everyone seemed to work for a federal agency, and where both my parents were involved with regulation, shaped my perceptions of science and policy. While I may be critical of some regulators' actions, I cannot say enough about the dedication, intelligence, and commitment of so many staff members in the federal agencies.

My editor at Yale University Press, Jean Thomson Black, has been a joy to work with. Susan Laity provided excellent suggestions to help clarify technical language. I thank the Press's anonymous readers for their thoughtful, detailed comments on the proposal and the manuscript. Their criticisms and suggestions improved this manuscript immensely. Any errors that remain are entirely my own.

Above all, I would like to thank my husband, Frank Goodman, for his love, encouragement, and support. You have created a wonderful refuge for us all along the Little Sugar River.

I thank *Environmental History* for permission to reprint selections in Chapter 3 that first appeared in "The Retreat from Precaution: Regulating Diethylstilbestrol (DES), Endocrine Disruptors, and Environmental Health," *Environmental History* 13, no. 4 (2008): 41–65; published by the

American Society for Environmental History and the Forest History Society. I thank Oxford University Press for permission to reprint material in Chapter 2 forthcoming in "New Chemical Bodies: Synthetic Chemicals, Regulation, and Health," *Oxford Handbook of Environmental History,* ed. Andrew Isenberg. Finally, I thank the University Press of Kansas for permission to reprint selections from my "Gender Transformed: Endocrine Disruptors in the Environment," in *Seeing Nature Through Gender,* ed. Virginia Scharff, 129–166 (2003).

CHAPTER 1

Disrupting Hormonal Signals

In March 2000, I joined an environmental justice field trip that met with women of Washington State's Shoalwater Bay Indian Tribe. One of the poorest tribes in the West, the Shoalwater were losing their tiny reservation to erosion and legal battles, and they were losing their future to a mysterious run of miscarriages. One woman after another described losing her fetus. They spoke to us of their grief, anger, sense of confusion, and fear that something in the water they drank or the fish they ate was killing their babies.[1]

The U.S. Centers for Disease Control and Prevention claims that the miscarriages could simply be random events, or possibly the result of genetic flaws. Or they could stem from diet, poverty, alcohol, or drug abuse, all of which can contribute to miscarriages. Few tribal members are reassured, for women on the reservation have taken meticulous care of their health during their pregnancies, yet they still have high rates of pregnancy loss. Many people in the tribe fear that the culprit could be environmental. Farmers spread pesticides on cranberry bogs near the reservation, foresters spray herbicides on surrounding forests, and oysterers use chemicals in Willapa Bay to control parasites that threaten the oyster industry. Many of these chemicals have the potential to disrupt the actions of hormones that shape fetal development. Yet because fetal development is so complex and because synthetic chemicals are so difficult to monitor, no one can determine exactly what is harming the developing children.

The expectant mothers of the Shoalwater tribe are being exposed to something, yet no one knows what. Their situation is extreme but not unique. Rich or poor, urban or rural, we are all breathing air that carries toxic dust from fertilizers, drinking water contaminated by plumes of toxins, eating food tainted with chemicals leached from plastic containers. The water that moves inside us is eventually the water that moves through the bodies of the Shoalwater women. It is the water that stagnates over the Superfund site behind my old house, and it is the water where fish swim, connecting one ecosystem to another, one species to another, and one body to another. Toxic chemicals have the potential to cross the boundaries between species and generations, altering the hormone systems that shape our internal ecosystems of health, as well as our relationships with the broader ecosystems around us.[2]

New technologies and methods for the detection of synthetic chemicals, particularly endocrine disruptors, have drawn increasing attention toward the pervasive presence of industrial chemicals in our bodies. In July 2005, the Centers for Disease Control released its *Third National Report on Human Exposure to Environmental Chemicals,* revealing that synthetic chemicals permeate bodies and ecosystems.[3] Many of these chemicals can interfere with the body's hormonal signaling system (called the endocrine system), and many are persistent, resisting the metabolic processes that bind and break down natural hormones.

In the 1980s the researcher Theo Colborn of the Conservation Foundation began documenting wildlife responses to pollutants in the Great Lakes. About one-fifth of U.S. industries and one-half of Canadian industries are located along the Great Lakes or tributary streams, making the region a microcosm for problems with pollutants in industrial society. Colborn found no shortage of wildlife problems in the area, but few consistent patterns. Some studies suggested elevated rates of cancer in certain species, others showed impaired fetal development, while still others found behavioral changes in wildlife.

Little seemed to tie these results together until Colborn learned of research by the biologist Frederick vom Saal showing that developing fetuses could be extraordinarily sensitive to tiny differences in fetal hormones. Vom Saal had noticed that female mice from the same litter showed dramatic differences in size and aggression. Because these mice were genetically identical, something other than genes was determining their differences. A female mouse's position in the womb turned out to powerfully influence her behavior when she reached adulthood. In the mother's uterus, females positioned next to their developing brothers were exposed to more androgens than those next to their sisters. In maturity, the mice located near their brothers were more aggressive and slower to mature, not because of genetic differences but because of tiny differences in prenatal hormones.[4]

Vom Saal's work made Colborn wonder whether the effects she was seeing in Great Lakes species might be linked to fetal development. If exposure to tiny doses of hormones could lead to significant effects later in life for laboratory animals, might the same be true for wildlife? Could synthetic chemicals be disrupting the endocrine system in developing fetuses? Colborn hypothesized that certain chemicals in the Great Lakes were mimicking estrogen, thus influencing the action of steroid hormones on fetal development, leading to reproductive problems in adulthood.[5]

The more researchers looked, the more they found that rivers and streams were laden with chemicals that had the potential to affect reproduction in wildlife. In the effluent of sewage plants, scientists found male carp and walleyes that were not making sperm but were instead producing high quantities of vitellogenin, an egg-yolk protein typically made by females. Other studies in the Great Lakes region found male white perch that had developed intersex characteristics. Students on a biology field trip in Florida noticed that every mosquitofish they found seemed to be a male, for each had a gonopodium—an anal fin that males use for copulation. But many of these apparent males turned out to be pregnant, and the students discovered that many of them were actually females that had developed gonopodia. As the biologist Mike Howell discovered, the problem was that wastes from pulp and paper mills were contaminated with chemicals that acted like testosterone. Around the world female killifish, sailfin mollys, blue-gill sunfish, American eels, and Swedish eelpouts had all become masculinized in streams that contained pulp-mill waste. Other fish species have become feminized by synthetic chemicals that mimic estrogen. In some western U.S. rivers, male Chinook salmon have developed female characteristics, while some male Atlantic cod and winter flounder have reduced testosterone levels, hampering reproduction.[6]

Sexual transformations were not limited to fish. Once researchers began looking, they found signs of reproductive problems in numerous species. Male alligators exposed to DDT in Florida's Lake Apopka developed penises that were one-half to one-third the typical size, too small to function. Two-thirds of male Florida panthers had cryptorchidism, a hormonally related condition in which the testes do not descend. Prothonotary warblers in Alabama, sea turtles in Georgia, and mink and otters around the Great Lakes all showed reproductive changes.[7] Male porpoises did not have enough testosterone to reproduce, while polar bears on the Arctic island of Svarlbard developed intersex characteristics. In one particularly disturbing example, Gerald A. LeBlanc of North Carolina State University in Raleigh found that more than a hundred species of marine snails were experiencing a condition known as imposex, a pollution-induced masculinization. Affected females could develop a malformed penis that blocked their release of eggs. Engorged by eggs that could not get out, many snails died.[8]

By the 1990s, researchers had noticed that not only wildlife species were showing difficulties with reproductive health; increasing numbers of people were as well. As with panthers, the incidence of cryptorchidism in British men has increased, doubling in two decades. Since 1970, boys in the United States have become increasingly likely to develop severe hypospadias, a birth defect of the penis. Testicular cancer has increased in many industrialized countries; in Denmark it has more than tripled since World War II, while in the United States incidence increased by 51 percent between 1973 and 1995. Similar increases have occurred in other Scandinavian countries and Scotland. Since the 1950s, sperm counts in some regions have declined significantly worldwide. Men in many industrial nations are showing increases in prostate cancer; a 1999 review found that men in the United States in 1994 had a much greater risk of being diagnosed with prostate cancer than their fathers had.[9] Much of the increase in the number of diagnosed cases is probably the result of better screening tests, but researchers are nonetheless concerned that actual incidence may also be increasing for unexplained reasons. Across the United States and Puerto Rico girls appear to be developing breasts at a younger age, and other signs of early puberty have also become apparent.[10] Epidemiological research on women's reproductive health has found an increase in the incidence of infertility, endometriosis, fibroids, breast cancer, and ovarian cancer since synthetic chemical production began to boom in the 1950s.[11]

What, if anything, connects all these problems with reproductive health? Many researchers now believe that these changes stem from disruptions of hormones by synthetic chemicals, particularly during vulnerable stages of fetal development. Hormones are chemical signals that regulate communication among cells and organs, orchestrating a complex process of fetal development that relies upon precise dosage and timing. Anything that scrambles the messages from hormone-signaling systems can alter patterns of development and health, just as scrambling airplane radio systems can alter flight patterns. The plane might not crash, but the static can disrupt the signals necessary for clear communication. The consequences may sometimes be minor, such as when the plane is in mid-flight at a steady altitude, but at other times — during take-off and landing, for instance — scrambled messages create havoc. Similarly, when synthetic chemicals alter hormone-signaling systems, adults might be resilient to the changes, but fetuses and young children can experience permanent transformations.[12]

Ever since endocrine-disrupting chemicals were first commercially produced in the 1940s, their hormonal mechanisms of action have posed novel challenges for scientists and regulatory agencies seeking to protect public health, because they do not easily fit within traditional risk paradigms. Toxicologists based their paradigms of risk on natural toxins that caused acute poisoning at high doses. As the environmental scientist John Peterson Myers writes, "Traditional toxicants are thought to work by starting a process (or stopping one) by overwhelming the body's defense system. Up to some level of contamination, the body can defend itself against chemical assaults." Chemicals that disrupt hormone systems act in a variety of ways, however, usually by changing signals that direct complex processes with intricate feedback loops.[13]

Even today, a popular Yale University Web site for poisons teaches that "the dose makes the poison." *All* toxins, this Web site states, are dose dependent: "The toxic effect of a substance increases as the exposure (or dose) to the susceptible biological system increases. For all chemicals there is a dose response curve, or a range of doses that result in a graded effect between the extremes of no effect and 100 percent response (toxic effect). All chemical substances will exhibit a toxic effect given a large enough dose. If the dose is low enough even a highly toxic substance will cease to cause a harmful effect."[14] Endocrine disruptors, however, violate every aspect of this definition of risk. Instead of being dose dependent, with a threshold below which the chemical is safe, endocrine disruptors typically demonstrate the following properties:

Dose: Their effects are often not dose dependent. Classic natural toxins such as poisonous mushrooms typically show a dose-response curve, with larger doses leading to more harmful effects than smaller doses (often in a linear relation: twice as much toxin leads to twice the effect). In contrast, endocrine disruptors may show greater effects at lower doses, depending on the timing of exposure rather than the dose alone.

Threshold: Natural toxins usually have a threshold of safety, or what is called a "no observable adverse effect level." At some point, for example, a sample of a poisonous mushroom will be so tiny that a person would not be harmed by it. In contrast, endocrine disruptors often lack this threshold. Even a single molecule diluted in a trillion molecules of water may have potential activity. These biological effects occur at doses that are orders of magnitude lower than current dose limits for other toxins.

Age: Effects often do not correlate to the size or weight of the exposed individual, as is usual with traditional toxins. A large person should be able to eat more of a poisonous mushroom than a small person before feeling the harmful effects, but the effects of endocrine disruptors are rarely so predictable. Age rather than size is often the critical factor. Infants and developing fetuses are most at risk, while adults can often show entirely different effects.

Timing: Endocrine disruptors often have effects that are not apparent immediately after exposure. Unlike natural toxins, which usually show effects almost at once, endocrine disruptors may not show effects for decades. A person who was exposed to synthetic endocrine disruptors such as DES in the womb might show no harm at birth but might develop cancer or reproductive problems at puberty.

Researchers detected many of these patterns in the 1930s and 1940s during their initial investigations of diethylstilbestrol. While many scientists believed that these unexpected patterns indicated a need for extra caution, industry advocates dismissed the possibility that the new chemicals might be causing harm because the observed effects violated standard beliefs about toxicology. Yet these unusual effects all derive from the ways hormones typically function.

A review of some basic principles about the body's hormone system (known as the endocrine system) can help us make sense of the ways synthetic chemicals can act as endocrine disruptors. These principles will frame my argument about why regulators struggled to respond to the risks of many synthetic chemicals. To explore them I shall focus on one group of hormones critical for sexual development in both males and females: the estrogens, which include estradiol, estrone, and estriol.

Estrogens are steroid hormones: that is, they are fat-soluble and derived from cholesterol.[15] Estrogen can be made in several locations within the body, but the ovaries are the most important production site in women of reproductive age. After the ovaries secrete estradiol, the molecules travel through the bloodstream until they encounter cells with specific receptor proteins that fit the hormone. Each hormone has a unique shape that fits the shape of particular receptor proteins at the target cell. Imagine the hormone as a key and the receptor protein as the lock. Only if the key fits can the door be unlocked. After estradiol binds to a matching receptor protein, it triggers a change in the shape of that protein, forming a new molecule called a hormone-receptor complex. The hormone-receptor complex enters the cell's nucleus and binds to its DNA, triggering a cascade of events in the cell, such as signaling the DNA to express particular genes, make particular proteins, or develop particular tissues. One familiar result is the instruction to breast cells to begin replicating during puberty. Even a tiny amount of estradiol that binds with the correct receptor can trigger the signaling cascade, with far-reaching effects such as breast growth.

Estrogen receptors are abundant in our bodies: in breast cells, the uterus, the ovaries, bone cells, hair cells, blood vessels, liver, kidneys, eyes, and even the prostate. Some hormone receptors for estrogens are unique, allowing only a single configuration of a molecule to fit. Other receptors are less specific, and many different chemicals can bind to them. A synthetic chemical that binds to an estrogen receptor might trigger cellular processes, effectively acting as an estrogen in the body. Other synthetic chemicals might bind to an estrogen receptor with antagonistic effects, blocking the binding of the body's own (endogenous) hormones. The PCBs in the Fox River, for example, can function as anti-estrogens by binding to a particular estrogen receptor and then preventing that receptor from binding to the body's endogenous estrogens.

While hormones are critical for life, too much of a given hormone can lead to havoc. Depending on timing, excess estrogens might stimulate the replication of cancer cells, signal tumors in a woman's uterus to grow, and transform patterns of sexual development. Because the levels of a particular hormone needed by a body can change from moment to moment, a complex suite of interconnected feedback systems governs hormone activity. This may regulate hormone synthesis within glands, control hormone release into the bloodstream, affect hormone uptake by target receptors, and alter the ways hormones bind to proteins so they can be broken down and removed from the body.[16]

Negative feedback systems function like a thermostat, maintaining homeostasis, or internal balance. When temperatures go up, the thermostat shuts the furnace off, and when the temperature drops low enough, the thermostat signals the furnace to turn back on. Similarly, when levels of the body's estrogens drop below a certain amount, an organ called the hypothalamus secretes gonadotropin-releasing hormone, which travels to another organ in the body (the anterior pituitary gland), which then secretes yet another hormone called follicle-stimulating hormone, which makes its way back to the ovaries and stimulates more estrogen production. Blood estrogen levels eventually rise high enough that the hypothalamus stops secreting its gonadotropin-releasing hormone, thus stopping the secretion of follicle-stimulating hormone from the pituitary gland, and that in turn stops the production of estrogen from the ovaries. Feedback systems potentially enable small amounts of hormones to create larger effects than high doses, because high doses can shut down hormone synthesis.[17]

Estrogen levels in the body are also regulated by serum-binding proteins known as sex-hormone-binding globulin. This protein binds with estrogen and other steroid hormones circulating in the bloodstream. Bound estrogens are unable to enter target cells, making these estrogens biologically inactive. When blood estrogen levels drop low enough, serum-binding proteins may release their estrogens, allowing them to become biologically active once again. Biologically active estrogen levels are thus determined not by estrogen production alone but also by the level of serum-binding proteins in the blood. Binding-protein levels depend on a complex balance of other chemicals known as enhancing and inhibiting factors. Hormones such as insulin may act as inhibiting factors, decreasing the level of serum-binding protein in the blood, and thereby increasing biologically active estrogen levels. Synthetic chemicals may do the same thing.

The complexities of the feedback, receptor, and binding-protein systems allow for rapid fine-tuning of estrogen levels in the body, but they also mean that synthetic chemicals can interfere with numerous different pathways.[18] Endocrine-disrupting chemicals may interfere with hormone signaling by altering the metabolism of steroid hormones or by inhibiting their synthesis. Synthetic chemicals may bind with serum-binding proteins so that those proteins cannot bind with the body's own endogenous estrogens, increasing their biological activity. Alternatively, synthetic chemicals may be much weaker estrogens than the body's own estrogens but be unable to bind with serum-binding proteins, making them biologically quite potent. This proved to be one of the major ways that DES and other synthetic endocrine disruptors affected the body. At certain times during pregnancy, estrogen levels increase dramatically, but the production of sex-hormone-binding globulin also increases, thereby protecting the fetus from the mother's high levels of circulating estrogens. DES is a weaker estrogen than the body's own estrogens, but it is less likely to be bound by serum-binding proteins, leaving the fetus vulnerable to the chemical's effects.[19]

Receptors and serum-binding proteins also influence the difference between natural and synthetic hormones. Beginning in the 1940s the livestock industry relied on what I call the "natural" argument to claim that their use of synthetic hormones in livestock was safe. When regulators and scientists raised concerns about synthetic estrogens, producers would point out that the body has high levels of its own natural estrogens and yet not everyone dies of cancer, so small amounts of a synthetic estrogen must also be safe. Because natural plant compounds in the human diet can also act as estrogens, industry advocates offered complex calculations purporting to show that natural estrogens were thousands of times more abundant than synthetic estrogens. If the body could survive high levels of natural estrogens, they argued, synthetic estrogens must also be safe, and regulatory staff were often persuaded. Many plants do indeed contain natural estrogenic compounds (called phytoestrogens), and when eaten in large concentrations, these may affect human reproduction. Like the body's own estrogens, however, phytoestrogens are quickly bound up by serum-binding proteins in the blood, and the body tends to flush them out rapidly. Synthetic chemicals, on the other hand, may be weaker estrogens, but because they avoid the chemical defenses of the woman's body, they can accumulate in body fat to toxic levels, persisting until pregnancy.

The effects of estrogenic chemicals such as DES puzzled researchers in the 1940s and 1950s because they differed dramatically among individuals, depending on the age of the individual and the timing of the exposure. These findings made little sense when interpreted through a standard toxicological paradigm, but they are less surprising when we consider how the endocrine system functions at different life stages. In adults, hormones mainly regulate ongoing physiological processes such as metabolism. Synthetic chemicals can lead to temporary endocrine changes, but adults are often able to recover from these disturbances. During fetal development, however, hormonal changes can have permanent, irreversible effects. Because a woman accumulates toxic chemicals over her entire lifetime of exposure, she can transfer much of her contaminant burden to her developing fetus during pregnancy, the time of greatest sensitivity.

Hormones orchestrate the complex dance of fetal development, telling various genes to turn on and off, and directing cellular replication and morphogenesis, the processes that transform simple collections of cells into complex organs. An embryo must develop from just two cells into an organism with trillions of cells and many organs, and hormonal signals guide the fetus through these developmental paths. Synthetic chemicals can disrupt critical steps, leading to effects that may become apparent only decades later, when the child reaches adulthood.

Early in life the endocrine system develops set points that control the number of hormone receptors and their sensitivity to changing hormonal signals throughout adulthood. When synthetic chemicals influence these hormonal set points in the fetus, the impacts are felt for a lifetime. Sexual development is particularly sensitive to these effects: for example, in the male fetus specialized cells known as Sertoli cells direct the development and descent of the testes, regulate the development of germ cells, and orchestrate the progress of cells that will secrete the hormones responsible for masculinization. Turning on too many estrogen receptors in the developing fetus can reduce the multiplication of Sertoli cells and fix their numbers at very low levels. This can affect descent of the testes and the development of urethral structures, setting into motion events that could lead to cancer decades later. Research on the developing prostate shows that exposure to synthetic estrogens such as diethylstilbestrol in the womb can lead to prostate problems later in life.[20]

Researchers in the 1940s and 1950s learned that exposing pregnant lab animals to synthetic chemicals such as DES could result in reproductive problems that emerged only at adulthood. Why, then, didn't they suspect that people might suffer similar effects? Some scientists were indeed concerned about the potential effects of synthetic chemicals on human reproduction. Yet after World War II, as genetic models of development began to dominate, few researchers remained interested in environmental influences on development. Genes rather than the environment were believed to set the blueprint for development.

Since the 1990s, an explosion of research in the field of epigenetics has transformed conceptual models of gene-environment effects on the developing fetus. Every cell in the body contains the individual's entire genetic code. But brain cells must use only the genes needed by the brain, while kidney cells should activate only the genes needed for renal function. Epigenetic processes direct how these different parts of the genome are activated or silenced during development. Cells commonly control gene behavior by attaching small molecules known as methyl groups to specific sections of DNA. The attachment and detachment of methyl groups is particularly important in the fetal development of the reproductive system, and hormones play key roles in these epigenetic processes.[21]

Exposure of the fetus to toxic chemicals can permanently reprogram tissue in a way that determines whether tumors will develop in adulthood. Many cells have tumor-suppressor genes that keep tumors from becoming malignant. Chemical exposure can lead to epigenetic changes that silence these genes, even when their DNA sequence is unchanged. Likewise, cells also contain tumor-promoter genes, which are normally suppressed. Exposure to synthetic chemicals can block the suppression of these genes, thereby allowing them to promote the growth of tumors. In animals bred to contain genes that make them particularly susceptible to fibroid tumors, those genes are normally suppressed, but exposure to toxic chemicals such as DES will turn those genes on in the fetus, and tumors will develop years later. Without the initial toxic exposure, however, such a genetic susceptibility may not lead to cancer in adulthood.

Development is no longer envisioned as an inevitable chain of events dictated by genes alone. Rather, developmental biologists now describe a complex symphony between cells, genes, organs, individuals, and environment, all influencing one another's melodies and harmonies. Genes may form the sheet music, but without the hormonal conductor to select which music to play and coordinate the musicians, cacophony would break out. Researchers in 2004, for example, exposed young mice to DES and observed epigenetic changes in the DNA that could cause the onset of cancerous growths in adulthood. Even quite low doses of DES altered methylation patterns and increased uterine tumor incidence, and these changes could pass from one generation to the next.[22]

Not all individuals respond in the same way to particular chemical exposures, making it difficult for epidemiological researchers to detect subtle effects. Experimental research on rats and mice, for example, shows that strains differ tremendously in their genetic susceptibility to endocrine disruptors. Although DES harms the rat thyroid, for example, strains differ in their sensitivity to DES-induced thyroid problems. In people, complex gene-environment interactions shape the likelihood of a woman getting breast cancer. Women with mutations in the BRCA1 and BRCA2 tumor-suppressor genes are more likely to get breast cancer, an indication of a genetic influence. But environmental factors influence whether these BRCA1 and BRCA2 genetic mutations will lead to cancer. For women born before 1940, before the boom in synthetic chemicals, having the BRCA1 and BRCA2 mutation has led to little increased risk of cancer. But for women born after 1940, having those mutations has meant a substantially increased cancer risk. These results suggest that environmental exposures can increase cancer risk even for women with an inherited cancer-susceptibility gene.[23]

As the women of the Shoalwater tribe have learned, the complex nature of hormone systems makes trying to connect any particular chemical exposure to particular reproductive problems extremely difficult. Pregnant women exposed to pesticides are more likely to have miscarriages, but this correlation is not firm proof that a pesticide caused the miscarriage. The case of PCBs illustrates some of the difficulties researchers encounter when they try to link chemical exposure to reproductive failure. PCBs are industrial chemicals that disrupt thyroid hormone function. The Environmental Protection Agency (EPA) banned their production in 1979, but Great Lakes fish still carry PCBs in their fat, and people who eat those fish accumulate the chemicals. Thyroid malfunction can lead to miscarriage, and because PCBs alter thyroid function, researchers suspect that PCBs might contribute to miscarriages. In lab studies, PCBs have been shown to change rates of thyroid-hormone synthesis, increase the metabolic clearance of some thyroid hormones, and cause miscarriages in rodents. Outside the lab, women who eat a lot of Great Lakes fish accumulate more PCBs than women who eat no Great Lakes fish. Yet it is not clear that eating such fish endangers women's fetuses. Women with a history of miscarriages do show higher PCB levels in their blood than women who have not had miscarriages, suggesting that PCBs may have been a contributing factor. But other studies have found that women who eat more fish from the Great Lakes do not have higher rates of miscarriages than women who eat less Great Lakes fish. Such findings, however, do not necessarily prove that low-level PCB exposure is safe. The PCBs in the fish may have had little effect on pregnancy; alternatively, the fish oils also present in the fish might have helped protect the developing fetus. Or perhaps the control women who were eating less Great Lakes fish were exposed to other synthetic chemicals that increased their miscarriage rates. Epidemiological correlations suggest paths for future research, but they rarely offer firm proof of either safety or harm.[24]

Laboratory studies show that other endocrine-disrupting chemicals can also lead to miscarriages. In rodents, experimental treatment with DES and bisphenol A (a chemical found in many plastics) both increase miscarriage rates. If the pregnant rodent manages to carry the offspring to term, the female offspring also show higher rates of miscarriages when they reach adulthood. A single chemical exposure, therefore, may affect three generations: the exposed mother, the developing daughter, and that daughter's potential offspring.[25]

In 2005 epidemiological studies on people showed that women with a history of recurrent miscarriage had higher levels of bisphenol A in their blood than women who had been able to carry their pregnancies to term. Yet the combination of epidemiological studies on people and experimental studies on laboratory animals does not provide proof that synthetic chemicals would cause the same effects in people that they do in other animals. Miscarriage, birth defects, and infertility have numerous potential causes for they are all part of a complex ecology of health. This complexity stems from the nature of endocrine systems, yet it has made political pressure against regulatory action difficult for federal agencies to withstand.

The womb is an environment of its own, yet one that is linked to the outside world. The chemicals that a woman has been exposed to throughout her life, not just what she consumes while she is pregnant, reach her fetus, connecting one generation to the next. Pregnant women hope that if they don't take drugs like DES, their children will be fine. But chemical contamination affects most women: 30 percent of pregnant women in one study had detectable levels of PCBs, DDT, and the pesticide lindane in their amniotic fluid, often at concentrations high enough to cause problems in lab animals.[26] What do these exposures actually mean for people? No one knows for certain, but a consensus is emerging that some synthetic chemicals—even at very low, background levels—can disrupt the signaling systems that shape fetal development.

Are low-level exposures to endocrine-disrupting chemicals a serious problem for people? Some of the central claims of the endocrine-disruption hypothesis are now agreed upon by all scientists, even within the chemical industry. Everyone agrees that wildlife exposed to certain synthetic chemicals show responses similar to those induced by steroid hormones. They agree that lab studies show that synthetic chemicals can bind with and activate hormone receptors, resulting in gene expression. They agree that exposing pregnant mice to extremely low concentrations of certain synthetic chemicals results in offspring with reproductive problems. They agree that some synthetic chemicals can make breast cancer cells multiply in cultures. They agree that persistent organic chemicals build up in human tissue and are passed to the developing fetus and the breast-feeding infant. They agree that many male fish and alligators exposed to industrial effluents show signs of feminization, a result also shown in the lab when eggs are exposed to some synthetic chemicals.[27]

But scientists still disagree on a fundamental issue: What do these animal and lab studies mean for people? Do people who do not experience occupational exposures have anything to worry about? Can endocrine disruptors explain any of the apparent increases in infertility, reproductive cancers, birth defects, reduced sperm counts, or lowered ages of puberty? Or are endocrine disruptors present at such low levels that they are a trivial concern?

In August 1999, the National Research Council of the National Academy of Sciences released its consensus report on endocrine disruption, commissioned in 1995 by the EPA and Congress. After four years of review and debates, the team finally managed to agree that endocrine disruptors at high concentrations do affect human and wildlife health, yet they could not agree on the extent of the harm caused by levels common in the environment. Moreover, the team argued that their disagreements were owing not only to gaps in scientific knowledge, but also to major epistemological differences on how to interpret the data and draw conclusions. The consensus report stated: "Much of the division among committee members appears to stem from different views of how we come to know what we know. How we understand the natural world and how we decide among conflicting hypotheses about the natural world is the province of epistemology. Committee members seemed to differ on some basic epistemological issues, which led to different interpretations and conclusions on the issues of hormonally active agents in the environment."[28]

The chemical industry's response to this report was to focus on the conclusion that no scientific certainty on human health effects had been established. Without certainty, the industry argued, endocrine disruption was not an issue for public health concern. As Myers writes, "This is a classic argument from industry spokespeople: that the absence of data proves safety. In reality, all it proves is ignorance." So, in the absence of firm proof, what should society do? Many in industry argue that we should do nothing until we have that proof. Others believe that such a course would be unethical, for as the Greater Boston Physicians for Social Responsibility stated, "We are engaged in a large global experiment. It involves widespread exposure of all species of plants and animals in diverse ecosystems to multiple manmade chemicals. . . . The limits of science and rigorous requirements for establishing causal proof often conspire with a perverse requirement for proving harm, rather than safety, to shape public policies which fail to ensure protection of public health and the environment."[29]

Epidemiological evidence is accumulating that supports the hypothesis that endocrine disruptors may be harming reproductive health, while experimental studies have found similar effects in laboratory animals. But researchers cannot ethically do these experiments on human fetuses to test whether the correlations between endocrine disruptors and reproductive disorders are real. Instead, regulatory agencies need to rely on the weight of the evidence from animal models and epidemiological studies, rather than experimental proof, to form policy.[30]

How can animal models be extrapolated to human effects? How can we understand the risks of low-dose, chronic exposures to synthetic chemicals? How can we understand effects on complex, interconnected systems? And politically, how can the government protect public health and the environment in the absence of complete proof? To understand the federal government's attempts to control the risks of synthetic chemicals, we need to explore early twentieth century debates about regulating the risks posed by natural toxins.

CHAPTER 2

Before World War II: Chemicals, Risk, and Regulation

Since World War II the production of synthetic chemicals has increased more than thirtyfold. The modern chemical industry, now a global enterprise of $2 trillion annually, is central to the world economy, generating millions of jobs and consuming vast quantities of energy and raw materials. Each year, more than seventy thousand different industrial chemicals are synthesized and sold, with the result that billions of pounds of chemicals annually make their way into our bodies and ecosystems.[1] Americans are saturated with industrial chemicals, the products of a post–World War II boom in synthetic chemical manufacturing.

Many of the reasons that agencies have had trouble regulating these new synthetic chemicals have to do with the risk frameworks that toxicologists developed before World War II to understand natural, but still toxic, pesticides such as arsenic. These frameworks were based on a set of assumptions about thresholds, impermeable bodies, and purity that worked reasonably well in addressing the effects of acute poisoning from natural toxins. But they have proven to be inadequate tools for dealing with the new threats posed by the synthetic chemicals of the postwar boom, particularly endocrine disruptors.

Most Americans take for granted their right to be protected from poisons, but in fact the government's power to regulate food and medicines was vigorously contested throughout the early decades of the twentieth century. Without legal compulsion, no manufacturers of chemicals were willing to undertake the time-consuming, expensive studies needed to establish the safety of their products, and this compulsion did not exist until the passage of the 1938 Food, Drug, and Cosmetic Act.

Farmers had been using inorganic poisons, particularly arsenic and lead, for two thousand years, but usually on a small scale. In the nineteenth century, as farmers worldwide converted ever-increasing swathes of native vegetation to cropland, a series of ecological changes meant that insects posed new risks. Fields planted with a single crop replaced native grasslands and forests containing a variety of species, and insect pest populations soared. The native predators of insects vanished along with their habitats, and the growth of global markets and trade networks allowed the new pests to make their way to new croplands. When the movement of Colorado potato beetles from their native wild vegetation in the Rockies to cultivated fields threatened to devastate potato farming in the eastern United States, farmers tried Paris green, an arsenic-containing pigment that protected potatoes from the beetle, despite early warnings that the poisonous effects might extend to human consumers. Impressed by the chemical's success with potatoes, farmers quickly extended its use to many more crops. Similarly, a gypsy moth outbreak in New England in the late nineteenth century brought lead arsenates into widespread use, for the lead mixture was gentler to foliage (if not to people) than arsenic. By the early 1900s, lead arsenate had become the most popular pesticide in use, and remained so until the introduction of DDT.[2]

Few regulations limited the use of pesticides in the early twentieth century, and few agencies questioned their risks. The federal government, in fact, encouraged their use as part of the modernization of agriculture. Farmers could use pesticides in whatever quantities they wished, and economic entomologists encouraged them to do so. Some farmers were reluctant to spray, worried about the expense or their family's health, but by 1900 professional entomologists seemed to have overcome any doubts they might have had about pesticide safety. *Farmers' Bulletin,* a publication of the U.S. Department of Agriculture, insisted that spraying arsenicals was "an operation virtually free from danger," and the department's chief entomologist, C. V. Riley, reassured readers about "how utterly groundless are any fears of injury."[3] By 1900 several states had acquired the regulatory power to *force* reluctant farmers to use pesticides.

Pesticides had come into use at a time when the risks of chronic exposure were difficult to measure or understand. Once researchers began to comprehend those risks, the pesticide industry was too firmly established to accept regulation readily, reacting passionately against suggestions that certain chemicals were neither necessary nor safe.[4] The effects of chronic, low-dose exposure to toxic chemicals became a key issue in early debates over the safety of arsenic in pesticides. When it became clear that arsenic residues remained on food, doctors, farmers, and entomologists began to argue about whether these residues posed risks. This debate set the parameters that still dominate current regulatory debates over toxic chemical exposure. Doctors believed that residues might be making farm workers ill, but agricultural experts argued that if a chemical did not lead to the immediate poisoning of the applicator and other farm workers, surely it could not harm consumers. Consumers, however, were coming into daily contact with substantial quantities of arsenic via "heroic" mineral medicines, pigments, paper, labels, toys, soap, preserves, dental fillings, money, and even wallpaper, whose arsenic-based pigments would slowly poison the air of the room it was in.[5]

Repeated crises during the late nineteenth century, in which children were poisoned by adulterated foods and contaminated medicines, sparked a consumer movement that pushed for regulation of both the chemical and food industries. Classic tests of environmental exposure were developed in response to these complaints. If people became ill in their homes, they were advised to relocate temporarily. If they felt better in the new environment but their symptoms recurred when they returned home, doctors judged the culprit to be environmental.

A group of doctors organized to appeal for legislation to control arsenic in consumer products, and in 1900 the first such act was passed in Massachusetts. Yet even doctors could rarely agree on whether low-dose arsenic exposure constituted a genuine threat to public health. At the same time, the possibility of slow intoxication from mercury, lead, and arsenic was being debated by toxicologists.[6] By the beginning of the twentieth century, in other words, debates over chronic effects from low-dose exposures versus acute poisoning from high-dose exposures had begun. These debates were to continue for another century.

One of the first regulatory efforts to assess risks to consumers from chemical exposure came with the passage of the Pure Food and Drug Act of 1906. After years of battling for the right to regulate food and drugs, the federal government finally obtained a limited power to protect public

Harvey Washington Wiley, the founder of the FDA
(Photograph courtesy of the FDA History Office)

safety from harmful chemicals. The chemist and pioneering consumer advocate Harvey Washington Wiley had led the fight against impure foods for decades, and his efforts eventually resulted in the founding of the Food and Drug Administration.[7]

Wiley, like other reformers in the Progressive era, was no enemy to industry. He believed that progress was essential for prosperity and that business was a critical driver of progress. Progressives also believed in the lessons of history: namely, that the nineteenth-century excesses of the robber barons had proven that business could not police itself. Business could not protect citizens from injustice and injury while also seeking profits. It was therefore the responsibility of government to monitor industry and protect the public.[8]

Wiley believed passionately in a form of the precautionary principle. Only precaution could protect the public from harm; waiting to see whether a particular chemical contaminant in food might be harmful was unethical, for it would turn America into a nation of guinea pigs. Wiley argued that because the early effects of chronic arsenic poisoning were almost undetectable, it would be impossible to regulate chemicals if proof of harm were required. He believed that what he called any "abnormal constituent of food," such as a preservative or pesticide residue, was likely to disturb the "natural, normal activity of the body's organs," and he further believed that pure food was therefore much healthier than food with any level of adulteration.[9]

Wiley's precautionary reasoning did not convince industry scientists, who demanded what the historian James Whorton calls "clear clinical signs of physiological damage" before they were willing to judge a substance harmful. These disagreements reflected both practical and ideological differences. Practically, the argument focused on where the burden of proof should lie. Wiley's precautionary principle held that a foreign substance should be presumed guilty until proven innocent; the industry believed the opposite. As Whorton writes, "One approach imposes a risk on the public, the other a hardship on business."[10]

The differences between Wiley and industry were not just practical, however; they also had conceptual roots in the history of toxicology. Wiley had distrusted the presence of any level, no matter how small, of a poison in food, but toxicologists were beginning to argue that substances generally had a threshold value below which the substance was unlikely to affect an individual's functioning or physiology. These beliefs about threshold values developed in a medical context that assumed bodies could be separated from environments. The historian Frederick Rowe Davis notes that "toxicologists, many of whom initially studied pharmacology, accepted the fundamental aphorism of toxicology, attributed to Paracelsus: 'The dose makes the poison.' . . . Such thinking permeates the science of toxicology."[11]

The origin of the word *threshold* hints at the implications of the threshold model of exposure. *Threshold* comes from the Old English *þrescold*, "doorsill, point of entry." Under the influence of germ theories of health, toxicologists came to believe that for harm to occur, a person needed to be exposed to an amount of the substance great enough to breach the barriers and cross the threshold of the body. A smaller amount, and the threshold would be inviolate, the separate home of the self unassailed. These assumptions colored the ways regulators and scientists approached potential toxic risks.

The defeat of epidemic diseases had been a major public health victory in the late nineteenth century, and germ theories of health had been critical to these victories. Yet as the historian Linda Nash argues in *Inescapable Ecologies,* germ theories had replaced earlier ecological ideas about health in which the body was envisioned as permeable to environments. When researchers gained insight into the microbial origins of many diseases, they began to see the body as separate from its environment, susceptible to penetration only after a germ or poison crossed the threshold of the body's barrier to the world. These beliefs helped medical scientists model the effects of acute poisoning, but they made it more difficult for researchers to imagine the new chemical threats that were arising with the growth of industrial agriculture and chemistry.[12]

During the 1920s, industrial hygienists developed techniques that, in Nash's words, enabled them "to quantify chemical exposures and to correlate those exposures with both physiologic variables and obvious signs of disease." The industrial hygienists envisioned chemicals "as akin to microbes, as singular agents that were capable of inducing a specific disease once they entered the body. What mattered was not the broader environment but the specific chemical exposure."[13] Researchers in industrial hygiene were well versed in physiology as well as germ theory, and, like physiologists, they believed in homeostasis, the concept that a body is a self-regulating system that can equilibrate itself, thus coming to balance again after being exposed to low levels of contamination. Wiley, trained as a chemist rather than a physiologist, grudgingly accepted the industrial hygienists' argument that chemical contamination might be inevitable in the modern world, yet he never agreed that bodies would equilibrate to this contamination.

Pesticide residues on fruits proved to be one of the first key political challenges for Wiley and his regulators as they struggled to apply the 1906 Pure Food and Drug Act in the face of concerted political opposition from the agricultural industry and its congressional allies. The central question was how best to establish the safety or harm of low levels of chemical residues on food. Where should the burden of proof lie? Given substantial scientific uncertainty over the potential effects of residues, how could regulators create good public policy?

One side, made up of farmers, their congressional allies, economic entomologists, the Public Health Service, and some medical scientists, believed that the criterion for safety should be clinical acute poisoning; doses that did not immediately make humans sick were assumed to be safe. The best way to test for toxicity would be to study people who were exposed to high levels of a toxic chemical; if they were healthy, surely the public would be safe. The burden of proof should be on those who wanted to regulate. The other group, including Wiley, consumer groups, and many medical scientists, took the opposite view. They believed that safety should not be defined by the absence of acute poisoning but rather by chronic effects, and that such data could best be gathered by extrapolating from laboratory studies conducted on animals over their entire lifetime. The burden of proof should rest on those who produced the chemicals, not the regulators.[14]

The political power of his opponents eventually thwarted Wiley's efforts at establishing a precautionary principle. In 1927 the Bureau of Chemistry (later to become the Food and Drug Administration) developed a "tolerance" policy based on the assumption that below a certain threshold a substance could not penetrate the body's defenses and become toxic. Harmful additives and toxic residues were permitted in foods up to certain quantitative limits called tolerances. These tolerances were quite liberal, based on assumptions that below the level that caused acute poisoning, a substance must be safe.[15]

Even with liberal tolerance standards, the Bureau of Chemistry would still be able to do little to enforce regulations in the early 1930s, for the country was entering the Great Depression, and growers were campaigning vigorously against any federal regulation. Manufacturers opposed strict interpretation of the act, and further opposition from within the Department of Agriculture made it hard for the bureau to regulate. The Bureau of Chemistry was part of the Department of Agriculture until 1940, and this created deep conflicts of interest since the department believed that its mission was to protect and advance American farmers, not regulate them. It opposed every effort of the Bureau of Chemistry to restrict chemical use.

Indeed, from its creation the bureau had struggled with more than the issue of threshold levels. The Pure Food and Drug Act of 1906 had held manufacturers accountable for the content of drugs, but it did not regulate either the drugs' safety or their effectiveness. In 1912, Congress had passed an amendment outlawing labels that made "false or fraudulent" claims, but this proved ineffective at protecting public health, for no one could prove that a manufacturer intended fraud. Wiley resigned that same year

Walter Campbell, the first FDA commissioner
(Photograph courtesy of the FDA History Office)

in frustration, leaving to join the editorial staff of *Good Housekeeping* magazine, where he could better advocate for consumers.[16]

By the 1920s, the medical community was beginning to discuss the dangers presented by synthetic chemicals in the industrial environment. Medical scientists such as Karl Vogel began to note that all people in industrialized nations carried traces of lead and arsenic in their bodies. Laboratory research showed that chronic exposure to arsenic created tissue degeneration in rats, and some physiologists, such as Sister Mary McNicholas, insisted that the same was likely to be true in humans. Industry and agricultural scientists rejected these arguments, claiming they were only hypothetical because they were not based on human experiments.[17]

All these debates confronted the first commissioner of the FDA, Walter Campbell. Campbell, a lawyer from Kentucky, had been hired by Wiley as head inspector in the Bureau of Chemistry soon after passage of the 1906 act. He remained after Wiley's resignation and became head of the bureau in 1921. In 1927, when the bureau's name was changed to the Food and

Drug Administration, he became its first commissioner. Campbell was a strong advocate for consumer health, but he was also a realist during the 1920s, a time when large corporations were gaining ever more power and regulators losing it. By 1929, two-thirds of the industrial wealth was controlled by large corporations rather than family businesses, and, as the historian Philip Hilts writes, "only two hundred corporations controlled half of American industry."[18]

As head of an agency with little power, Campbell tried to develop what he called a partnership model with industry. Drug sales had increased dramatically since the Pure Food and Drug Act was passed, but consumers still lacked protection from unscrupulous or careless manufacturers. Instead of forcing businesses to take the necessary steps to protect the public, Campbell believed that the FDA had no choice but to form partnerships in which the administration would first attempt to educate industry about potential problems, then alert them to violations, and finally negotiate with them to stop the worst abuses. Wiley complained bitterly when Campbell adopted this strategy, but Campbell felt he had little power to do otherwise.[19]

Pesticides were particularly contentious for the new FDA. Many new pesticides were being placed on the market, and many new residues were finding their way into processed food. But the 1906 act gave Campbell's agency no power to even find out what was being used on food, much less the ability to control the pesticides. The FDA had a difficult time enforcing even limited seizures of adulterated food or food contaminated with toxic residues because growers were willing to contest the agency in court. Fruit growers in particular insisted that the FDA should not release information about excessive residues, arguing that members of the public would panic. The industry urged the agency to keep findings confidential, giving growers a chance to fix the problem.

In 1929, Campbell warned apple growers that the strategy of hiding information from consumers posed grave risks to business: "What do you suppose would happen if the general public became acquainted with the fact that apples were likely to be contaminated with arsenic? . . . So far, we have not given the matter any publicity, and the public as a whole has no general knowledge on the subject. . . . If they were to become curious today, we would probably have to admit that we are perhaps not doing everything possible to remove excess arsenic from our fruit."[20] Campbell was correct. When consumers learned how little was being done to protect them from pesticide residues, outrage did indeed result. The consumer protection movement that emerged during the 1930s attacked the comfortable secrecy that had developed within government agencies, helping to catalyze regulation of pesticides and other chemical toxins.

The New Deal era of the 1930s transformed American attitudes toward regulation. After Franklin Delano Roosevelt became president in 1933, his administration began the long process of regulatory reform. One of Roosevelt's advisers was Rexford G. Tugwell, an economics professor who became second in command at the Department of Agriculture. The state of food laws in America horrified Tugwell. Convinced that the FDA had been, as he wrote in his diaries, "perverted by the attempt to protect business interests," Tugwell was determined to change the agency.[21] Immediately after the president's inauguration, Tugwell began lobbying for reform of the 1906 Pure Food and Drug Act. Campbell and Tugwell convinced the new president of the need for a major revision in the food and drug laws, and with Roosevelt's support, the process of reform began. The agency finally broke what Whorton called "its tradition of sheltering the residue problem in secrecy."[22]

Even with growing consumer support, reform proved difficult, however, because fruit growers and chemical companies were able to fight off FDA attempts to restrict residues with help from their congressional representatives. In one key episode with lasting repercussions for chemical regulations, in 1935 Representative Clarence Cannon of Missouri censored Campbell's testimony about the health effects of residues. Two years later, when the FDA used experimental evidence from animal studies showing health problems connected with pesticides, Cannon made certain that the appropriations act for fiscal 1937 included a provision that "no part of the funds appropriated by this act shall be used for laboratory investigations to determine the possible harmful effects on human beings of spray insecticides on fruits and vegetables."[23] Politics, not science, in other words, created an intense opposition to using animal experiments as a basis for regulating chemical risk—an opposition that continues to this day.

Although the effort took five years, the reformers finally got the Food, Drug, and Cosmetic Act passed in 1938. This act, while not as strict as many consumer advocates had hoped, was a major victory for the administration and for the concept of precaution in drug regulation. The law required that before a drug was placed on the market, manufacturers had to submit to the FDA what were called New Drug Applications. These had to include information about the content of drugs, the manufacturing process, their intended uses, and safety evidence. The government had sixty days (which could be extended to six months) to raise objections to the New Drug Applications; if the government failed to respond within that time period, the drug was automatically approved.[24]

Even with passage of the 1938 act, unresolved tensions remained around questions of risk. While drugs would be closely regulated and subject to precautionary measures, pesticide residues and household and agricultural chemicals would not be. The role of animal experiments in assessing drugs was still contentious, particularly now that the funding of investigations into the risks of pesticides had been outlawed. Disputes over the risks of low-level, chronic exposure to synthetic chemicals had begun, but without resolution. Conceptual models of health and the body were still in flux. While precaution was a central element of the 1938 act, consumers and regulators still lacked the power to ensure that the FDA would be able to protect public health in the face of political and legal challenges, even after the agency gained independence from the Department of Agriculture. In the 1940s, contentious battles over approval of the first synthetic estrogen, diethylstilbestrol, would test the FDA's resolve.

CHAPTER 3

Help for Women over Forty

In 1941 an article by the journalist Helen Haberman in the *Reader's Digest* advised women that "help for women over forty" might soon be available. Nine million subscribers to the magazine learned that a wonderful new drug could relieve aging women of "the most distressing of natural body processes," but only if the FDA were willing to approve the new synthetic estrogen, diethylstilbestrol. As the article explained, "Called by one clinician 'the most valuable addition to our therapy in recent years,' the first synthetic estrogen, diethylstilbestrol . . . awaits only the approval of the Federal Food and Drug Administration as a 'new drug' before being placed on the market as an inexpensive, orally-administered therapeutic agent for the relief of the menopause and other conditions."[1]

Haberman noted that "many [women]—perhaps 40 percent—suffer pain and discomfort, with characteristic sensations of heat and cold known as 'flashes.' Crying spells, sleeplessness, and just 'nerves' are other symptoms. Until recently the majority of these women suffered in silence. Many feared insanity, for depression wasn't unusual. Uninformed or inhibited, they did not consult a doctor." Haberman was not the only journalist of her era to view menopause as a crisis and a tragedy, a time of potential madness when women lost what was most essentially female about themselves: their reproductive potential. What was unusual about the article was the solution it proposed. A "sensational" new drug could provide "inexpensive relief for a dreaded crisis of discomfort and depression."[2] Soon after the article appeared, letters flooded into FDA offices from women begging that diethylstilbestrol be made available. Some even wrote to President Roosevelt, asking him to weigh in on the drug-approval process, and his staff did indeed communicate with FDA staff about the hormone.

This episode illustrates a key transformation in Americans' conception of women's health. In the late 1930s, menopause had increasingly become defined as a condition in need of treatment, rather than a normal part of aging. The year after the Food, Drug, and Cosmetic Act was passed, drug companies proposed that the newly synthesized chemical diethylstilbestrol be approved for the treatment of menopause. As a new drug, DES would be subject to the provisions of the 1938 act, which required that the drug companies provide detailed information about its proposed uses and evidence of its safety.

Diethylstilbestrol quickly became an important test case of the government's authority to regulate new drugs. While debating whether to allow DES to be prescribed to women, federal researchers, regulators, and pharmaceutical companies all learned that it caused cancer and altered sexual development in fetuses. Why, then, did the FDA approve the use of DES in 1941? The development of DES as a treatment for menopause symptoms occurred as hormone research was dramatically changing understandings of sexual differentiation. Scientists learned in the 1930s that ovaries produced estrogen and other hormones that led to the development of female sexual characteristics in women. According to this formulation, females produced female sex hormones such as estrogen, which made them feminine, and males produced male sex hormones such as testosterone, making them masculine. But this simple idea soon became more complex. The historian of medicine Nellie Oudshoorn writes that when endocrinologists realized that both sexes contained both male and female hormones, the "shift in conceptualization led to a drastic break with the dualistic cultural notion of masculinity and femininity that had existed for centuries." This transformed biological definitions of sex, Oudshoorn notes, for "the model suggested that, chemically speaking, all organisms are both male and female. . . . In this model, an anatomical male could possess feminine characteristics controlled by female sex hormones, while an anatomical female could have masculine characteristics regulated by male sex hormones." A simple "theory of duality (sex difference) was transformed through the bio-chemists' challenge."[3] As medical technologies for seeing into the body developed, the meaning of sex became less defined by characteristics visible to the naked eye, such as the penis and breasts, and increasingly by characteristics such as hormones that were hidden from the ordinary gaze.

With the explosion of hormone research in the 1930s, scientists began to understand that hormones alone did not define sexual differences, for males as well as females were influenced by estrogen. Yet the belief that sexual difference was fundamental did not vanish. The location of sex difference simply moved from the gonads to the whole body. This shift helped create the concept of the hormonal body in which hormones ruled women's nature, and medical intervention during menopause might transform an unruly, disordered hormonal environment into a regulated order.[4]

Scientists constructed a model that they believed explained the decline in women's hormones, particularly estrogen, as women aged. Yet actual women often failed to fit into that neat model. During perimenopause, hormones do not simply decline; they fluctuate unpredictably, and these changes can provoke a variety of symptoms. To make women's bodies controllable and predictable, doctors and scientists joined forces. In the last three decades of the nineteenth century, many surgeons had advocated removing women's ovaries when they experienced menstrual problems or reached their forties, past childbearing age. Following what became known as Battey's operation, named after the American surgeon who developed the procedure, doctors could then replace the natural estrogens with precise, regulated levels of hormones, transforming an unpredictable variation into an orderly process.[5] Surgery and hormone synthesis had appeared to give doctors the tools to regulate the internal ecologies of female hormone systems.

By the 1930s, doctors had realized that they did not need to remove the ovaries; instead, they could simply give women ovarian extracts to rationalize the variations in their aging bodies. The problem was that biological hormones were expensive, short-acting, and often ineffective when taken orally because metabolic processes quickly broke them down. Injected estrogens were more effective, but they were expensive and painful. All this meant that relatively few women underwent estrogen-replacement therapy in the 1930s. Nevertheless, some doctors promoted the therapy, not just to treat symptoms of aging but to arrest the aging process itself. Clinical doctors, in particular, were fascinated by the possible curative and restorative powers of the so-called glandular extracts.[6]

While clinicians and scientists debated whether restoring the sexuality of older women was advisable, both agreed that alleviating symptoms of menopause such as hot flashes, nervous tension, and interrupted sleep would be worthwhile.[7] Scientists and clinical practitioners differed in their training and perspectives, yet they agreed that medical experts should be managing female patients throughout menopause. Menopause became defined as a "deficiency disease" in need of treatment, and women's health became increasingly controlled by professionals. Doctors, researchers, and pharmaceutical companies developed networks of expertise to develop and market new drugs for the treatment of menopausal symptoms. It was in this context that the search for a cheap, synthetic estrogen intensified during the late 1930s.

Cultural assumptions about the nature of women and their unruly bodies framed the beliefs about menopause in the medical and scientific communities. Yet as the medical historian Judith Houck argues in *Hot and Bothered: Women, Medicine, and Menopause in Modern America,* women were not mere pawns of medical experts; they were often eager to change their experience of aging with new drugs. The synthesis, federal approval, and marketing of DES were part of a movement to medicalize women's health, but as consumers and patients, and later as advocates for research, women were key participants in the conversations about synthetic estrogens. As the historians Andrea Tone and Elizabeth Watkins write in *Medicating Modern America: Prescription Drugs in History,* new drugs can change how we view the boundaries between health and illness, and hormone-replacement therapy certainly changed the ways that Americans viewed aging.[8] Menopause was becoming a condition that could be treated with the help of new partnerships between doctors, drug companies, researchers, and government agencies, and the synthesis of diethylstilbestrol in 1938 by the British biochemist Edward Charles Dodds would prove critical to that transformation.

Throughout the 1930s, Dodds and the researchers in his laboratory were fascinated by the close structural similarities between estrone (a form of estrogen in women's bodies) and a group of carcinogens known as the phenanthrene group. Curious about potential links between carcinogens and estrogens, Dodds and his co-workers found that the phenanthrenes were estrogenic in lab tests, stimulating the growth of mouse uteri in the same way estrogens did.[9] Dodds soon realized that a molecule did not have to look like a natural estrogen to have estrogenic activity. A series of papers from Dodds's laboratory beginning in January 1934 focused on the estrogenic activity of phenanthrenes and phenol compounds. Dodds then turned to other compounds of similar structure, and in February 1938 his team published an article in *Nature* announcing that the compound his laboratory had recently synthesized, diethylstilbestrol, was extraordinarily estrogenic. Hormonal functions in humans and other mammals could be induced by synthetic substances, something no one had been certain was possible.[10]

After the discovery of diethylstilbestrol (which Dodds never patented), manufacturers were able to synthesize DES quite cheaply from coal-tar derivatives. In 1939, a year after the DES results were published, U.S. pharmaceutical companies submitted New Drug Applications to the FDA for approval of DES to treat the symptoms of menopause. Two years later the drug was approved, over intense opposition within the agency. The FDA had only obtained the authority to require that drugs be demonstrated to be safe in 1938, and DES was the agency's first controversial test of the new law. Regulators had good reason to be cautious about how they proceeded because little consensus existed concerning the right of the federal government to regulate industry in the name of public health.[11]

Many pharmaceutical companies that are now being sued by the children and grandchildren of women who took DES have argued that the chemical was approved because nobody suspected it might be unsafe. The financial implications of this argument are significant, but the archival evidence refutes it. Before approving DES, the FDA requested that the leading pharmaceutical companies applying for approval collate all the European and U.S. studies on the chemical.[12] The document, an annotated bibliography prepared by the pharmaceutical companies, shows that they were fully aware of the research described within it.

Within a few months after Dodds's initial reports, researchers had begun publishing studies showing the harmful effects of DES on reproductive organs, sexual development, and fertility. In September 1938, Dodds's own laboratory reported that DES could prevent fertilized eggs from implanting in the uterus of rabbits and was also "highly effective in interrupting established pregnancy in rabbits." A month later, Dodds's group found that DES led to "atrophy of the testes, prostate and seminal vesicles in the male, ovarian atrophy in the female, and an increase in the weight of the adrenals and pituitary in both sexes." DES also atrophied testes in adult male rats, while retarding the development of testes in immature rats.[13]

When researchers fed DES to pregnant rats, mice, and chickens, the chemical led to changes in sexual differentiation in their developing offspring. Many of these deformities were not present at birth but emerged only when the offspring reached the age of sexual maturity. Researchers were particularly intrigued by the ways in which DES altered sexual differentiation during fetal development, leading to the development of intersexual traits. The testes appeared normal at birth, but when the offspring reached puberty their testes failed to descend (a condition called developed cryptorchidism). The chemical modified sexual development in the female fetus as well. When female offspring of lab subjects that were fed DES during pregnancy reached adulthood, their vaginas and urethras had a common orifice, and the ovaries were small, foreshadowing the high rate of fertility problems in male and female DES children. These studies provided a compelling suggestion that developmental problems stemming from synthetic hormone exposure might not be apparent until sexual development was under way.[14]

The potential for DES to cause cancer was a particular reason for concern. The drug, after all, had emerged during a larger debate in the 1930s about the potentially carcinogenic effects of estrogens, including natural estrogens. Like many endocrinologists of the era, Dodds recognized a key similarity between estrogens and synthetic carcinogens: both made cells replicate rapidly. In 1933, Dodd wrote to the journal *Nature* that "because cell proliferation which characterizes the estrus state is in some respects reminiscent of the early stages of a malignant growth, we have sought a correlation between substances having estrogenic action and those having carcinogenic properties." By 1939, when the application for DES was submitted to the FDA, nearly all researchers agreed that natural estrogens had the potential to be carcinogenic in laboratory animals, and that DES was at least as carcinogenic, if not more so, than natural estrogens because it was more potent at what researchers called exciting estrogenic effects. Researchers also found experimental evidence that DES might alter thyroid function, leading to thyroid cancer.[15]

What cancer studies in laboratory animals meant for women was uncertain, however. Experimental biologists tended to argue that animal studies on cancer initiation suggested that women would respond in similar ways. Practicing physicians, on the other hand, tended to dismiss animal studies, arguing that if the women they treated with estrogens did not immediately develop cancer, estrogens must be safe, even over the long term. Yet when physicians treated women with supplemental estrogens and some of those women eventually did develop cancer, the same doctors dismissed as mere hysteria the patients' conviction that the estrogens had caused their cancers.[16]

Researchers and clinicians debated about why some women exposed to estrogens developed cancer and other women did not. Individual genetic variation in susceptibility to toxic chemicals emerged as a critical area of research only in the 1990s; during the 1930s and 1940s, researchers' mechanistic models of how bodies worked did not offer much insight into how individual variations in cancer might arise. If the body's own hormones could be carcinogenic, why did not all women get cancer? Why would some women be protected but not others? While clinicians tended to dismiss the possibility that a natural substance produced by the body could induce cancer, FDA staff corresponded with researchers who believed that while natural estrogens might well be carcinogenic, most women's bodies were able to detoxify their potentially harmful effects. The researchers worried that those mechanisms might not work against synthetic estrogens such as DES, making the synthetic chemicals more carcinogenic than the natural substances.[17]

By December 1939 enough researchers were troubled by DES's ability to induce cancer that the *Journal of the American Medical Association* published an editorial titled "Estrogen Therapy—A Warning." The editorial argued, "The possibility of carcinoma cannot be ignored. . . . It appears likely that the medical profession may be importuned to prescribe to patients large doses of high potency estrogens, such as stilbestrol [DES], because of the ease of administration of these products." And the association's Council on Pharmacy and Chemistry warned that "the carcinogenic aspects of naturally occurring estrogens when injected into mice of a carcinoma strain [are] well known." Because of these concerns, DES "should not be recognized for general use or for inclusion in New and Nonofficial Remedies at the present time and . . . its use by the general medical profession should not be undertaken until further studies have led to a better understanding."[18]

FDA regulators in 1939 paid close attention to the Council on Pharmacy and Chemistry's warnings. Staff were quite troubled by laboratory research findings, even though little consensus existed on their applicability to humans.[19] The New Drug chief, James Durrett, was particularly suspicious of DES. Beginning in 1939, immediately after the New Drug Applications for DES came into his office, Durrett visited numerous scientists to gather their opinions about the chemical, and he sent a squadron of FDA bureaucrats around the country to interview clinicians familiar with estrogen treatments.

Durrett found no shortage of scientists concerned about supplemental estrogen, and he collected extensive interviews with them, using these to challenge industry claims of DES's safety. Each time a drug company said DES was less toxic than natural estrogens, Durrett found a researcher who would offer laboratory evidence to refute that claim. Each time a drug company neglected to mention a study showing toxicity, Durrett wrote to the head of the company with a reminder that such studies existed and asking why they were not included in the New Drug Application. Each time a drug company submitted studies performed by a small subset of researchers who liked DES, Durrett pointed out that other researchers had a very different perspective. Each time an industry representative insisted that animal studies had little meaning for human subjects, Durrett found scientists who disagreed.

One researcher, for example, R. Kurzrock, sent Durrett a manuscript in September 1939 summarizing his clinical findings in humans and showing that 80 percent of patients treated with DES developed toxic symptoms, including nausea and vomiting. Kurzrock was particularly concerned that he had observed "no relation between the size of the dose and the development of toxic symptoms." This was a red flag for researchers because it violated the core "dose makes the poison" precept of toxicology, suggesting that a safe level of exposure could not be achieved merely by reducing the dosage. Kurzrock noted that "there is no evidence of an acquired tolerance to the drug," another sign that DES might have unpredictable effects.[20]

Another researcher, Dr. Ephraim Shorr, wrote to Durrett that "the use of this drug in the human is associated in a high percentage of cases with very distressing by-effects in the form of nausea, vomiting, diarrhea, drowsiness and lethargy. . . . The reactions may occur in any dosage. They are frequently alarming and prostrating." Nausea and vomiting among women treated with DES varied tremendously from study to study, however. Shorr was frustrated when other researchers failed to reproduce his results, complaining in a letter to Durrett that these other investigators had "not observed their patients with sufficient diligence," ignoring women's complaints and failing to report symptoms. Shorr was also deeply concerned about long-term effects because he knew that women might be prescribed the drug for "each day for several years," and it was impossible to tell from short-term studies whether "liver damage and nerve damage would . . . result from taking this drug" over the long term. An absence of short-term data showing liver toxicity did not prove that the chemical was safe over the long term, Shorr insisted, and Durrett agreed with him.[21]

The nausea and vomiting that Kurzrock and Shorr had observed suggested to Durrett that DES might affect the liver, yet liver enzyme studies did not always show measurable changes in women who took the drug. Researchers did not know whether the inconclusive findings meant that no important effects were occurring or whether the technology available was simply not accurate enough to detect important changes. The report of one researcher, Dr. Engle, noted that "since liver damage tests are notoriously inaccurate they felt that they might actually be encountering some liver damage which their study would not detect. . . . They had developed no evidence of liver damage, but on the other hand they developed no facts which would justify the conclusion that liver damage was not taking place." Drug companies interpreted a lack of evidence of damage as evidence of safety, but the FDA staff agreed with Engle that such negative findings might merely indicate technological limitations in detecting harm, rather than a true absence of harm.[22]

Metabolism emerged as a central concern for researchers on DES. Put simply, no one understood how DES affected the body's metabolic processes or how the body broke down the drug. All scientists knew was that the synthetic estrogen was metabolized in different ways from those of the body's natural estrogens. Many researchers were perturbed by the observation that women's bodies did not quickly break down DES, as they did the natural estrogens, which were rapidly metabolized and excreted. Two researchers, Bernhard Zondek and Felix Sulman, noted that "in contrast to oestrone, stilboestrol is only rendered inactive in the organism to a small extent. . . . The fact that the organism is unable to inactivate considerable amounts of stilboestrol probably helps to explain its eventual toxic activity (compared with oestrone) particularly if large doses are used." Additional research showed that DES remained estrogenic even when excreted from the body: the feces from treated experimental animals could induce uterine growth in mice.[23]

DES also appeared to change the body's metabolism and growth, and researchers thought this might be due to effects on thyroid hormones and calcium metabolism. In a letter to William Stoner of Schering Corporation in April 1939, Durrett requested any references Stoner might come across showing reactions to DES. Stoner responded by pointing out that DES affected calcium metabolism and therefore altered growth. The same was true of natural estrogens, but as Stoner noted, after natural estrogens were discontinued, normal bone growth resumed; when DES was discontinued, bone growth failed to return to normal, showing that "irreversible bone changes are produced by stilbestrol." Stoner added that "the mammalian organism has become accustomed to the action of certain hormones which may produce damage although they usually do not, while [synthetic hormones like DES] . . . more or less uniformly cause damage with which the organism has not learned to cope." Stoner reasoned that natural estrogens might sometimes be harmful, but synthetic estrogens presented novel challenges to the body, making them more dangerous than the body's own estrogens.[24]

Stoner was not an unbiased observer. He worked for Schering Corporation, the German pharmaceutical company that had patented one of the natural forms of estrogen, so his motives in disparaging a synthetic estrogen may not have been entirely impartial. Similarly, researchers from Parke Davis, an American corporation that sold a natural form of estrogen, did their best to stress the dangers of synthetic estrogens that might compete with their own product. Nevertheless, eager as these researchers may have been to point out the advantages of the natural estrogen sold by their firms, their concerns proved valid. The ability of synthetic hormones and hormone mimics to bypass the body's own feedback systems and ecologies of health would eventually turn out to be a critical issue for drug regulators.[25]

The drug companies promoting DES argued that because the body naturally produces a great deal of estrogen during pregnancy, additional estrogens would necessarily be safe if the levels remained lower than those of the body's own sources. As early as the 1930s, however, researchers knew that estrogens do not act in linear or predicable ways. Sometimes the body produces a great deal of estrogen, and at certain times of a woman's life that surge in estrogen can protect against future cancers. At other times, however, a tiny bit of additional estrogen can promote cancer. There is no single safe level, and no reason to assume that additional levels will be safe as well.

One of the findings that most puzzled researchers in the 1930s and 1940s was that low doses of DES could be more toxic than high doses, violating the toxicological principle that the dose makes the poison. Pregnant patients received far more DES than menopausal patients but were rarely made nauseous by the drug, while menopausal women on much lower doses of DES were often nauseated. This suggested one of two things: either the effects of estrogenic substances were not dependent on dosage or researchers had been careless in their experiments. In September 1939, Kurzrock was surprised that he could not find any correlation between dosage and toxicity, and that no dose seemed small enough to ensure the absence of a toxic reaction. Kurzrock's unusual result led many researchers to dismiss his studies, even when other researchers confirmed them with reports of similar responses in their patients. But Jack Curtis, the chief pharmacologist of the FDA, was concerned enough by Kurzrock's results to urge that long-term studies in primates be done to explore potential carcinogenicity before the drug became available for human use.[26]

Although Curtis urged that safety studies be done on primates before doctors prescribed DES to women, little consensus existed about the applicability of animal studies for human subjects. If a chemical induced cancer in laboratory animals, did that mean it would do the same in women? No one was certain. One problem was that rodents, the most common lab subjects, had a much shorter lifespan than humans. In the course of a typical experiment, researchers could observe the induction of cancer. Yet in that same passage of time, clinical trials on humans rarely revealed cancer formation. Did this mean that the substance did not induce cancer? Many clinicians assumed that if they gave a drug to women for several years and no cancer developed, the drug was safe, because rat cancers typically emerged within that time. Other researchers pointed out that because humans lived much longer than lab rats, cancer induction as well might take much longer.[27]

Clinicians disagreed with laboratory researchers on the applicability of animal studies, and these disagreements sharpened over DES. Many clinicians did not believe that experimental studies on DES in lab animals had meaning for people. Although lab studies on rodents were rarely intuitively convincing to clinicians, personal experience often was. Clinicians tended to dismiss experimental data and argue from their own experience. For many clinicians who treated women with DES, if they failed to see problems shortly after treatment, DES was safe. Durrett wrote to Shorr that an appraisal of the safety of DES "apparently can only be satisfactorily approached by careful study in human beings. . . . A great many of our outstanding workers in this field are exclusively dealing with animal experimentation . . . and very infrequently do you see a worthwhile appraisal where the castrated female human has been used in numbers sufficient to warrant conclusions."[28]

The Council on Pharmacy and Chemistry of the American Medical Association (AMA) reviewed the research on estrogens in laboratory animals and concluded that all estrogens had the potential to cause cancer in women, even though data proving that estrogens caused cancer in women were not available. Absence of such data should not be taken to mean cancer would not be promoted, the council warned, given the strong animal evidence and the reported clinical effects on women. In response to this warning, the FDA sent out a press release in January 1940 to *Science News Letter*, aimed at clinicians and researchers, stating that stilbestrol was "effective but potentially dangerous" and citing the AMA and the Council on Pharmacy and Chemistry as authorities. The FDA concluded with a strong note of precaution: "Liver damage and cancer are among the possible dangers seen in use of the new synthetic hormone. The medical profession in general is advised not to use it until further studies have been made by experts."[29]

DES presented a particular challenge to the FDA, because as Watkins argues, "it was the first drug to be reviewed that did not purport to cure a disease yet did have the potential to harm users. Thus regulators took seriously the stipulation that the manufacturers provide sufficient evidence of DES's safety."[30] In conversations with FDA staffers, several cancer researchers urged the FDA to adopt a principle of precaution given the uncertainties that remained about DES and the relation between the drug's potential risks and its benefits. One physician, I. Penchars, told FDA staff who came to interview him about estrogens that because of the lack of understanding about cancer, precaution was the only realistic approach. Penchars was referring to estrogens used in face creams, not estrogens used for menopause, but the principle was the same. Patients should have the right to knowingly assume the risks posed by new drugs when the benefits outweighed the harm of an illness such as cancer. But with estrogens used in cosmetics and menopause, women were assuming unknown, possibly substantial, risks from chronic exposure to treat a normal process of aging.[31]

In memos written during 1939 and 1940, Durrett and Walter Campbell, the commissioner of the FDA, agreed that they were obliged to follow what they called "the conservative principle": if evidence were not available that showed clear absence of harm, a prudent regulator would assume harm might exist. When one DES manufacturer accused Campbell in 1940 of being "unscientific" for slowing the approval of DES within the FDA, Campbell wrote back: "In your letter of July 23 you indicate that one or more physicians do not believe that there is definite proof that estrogenic substances will cause cancer. We are aware that all physicians are not equally impressed with the evidence that estrogenic substances may induce cancer under certain conditions. However, in view of the very serious and often fatal nature of cancer, we believe that a conservative view point on this question is wholly warranted."[32]

Given the absence of strong evidence that DES was safe, and given the scientific uncertainty over its mechanisms of action and metabolism, Durrett urged Campbell to refuse to approve it, not because he had any proof that the drug would harm women, but because he had no proof the drug would *not* harm them. Campbell followed Durrett's advice, and in 1940 the FDA told the companies to withdraw their New Drug Applications for the approval of DES. Campbell noted that this decision was not final, and the drug companies would be allowed to resubmit if they could gather sufficient evidence showing DES to be safe in women.[33]

The pharmaceutical companies withdrew their individual applications from the FDA and decided to pool their resources, forming a group known as the Small Committee that would attempt to reverse the FDA rejection of DES. The Small Committee, led by Don Carlos Hines, who worked for Lilly, represented a group of pharmaceutical companies interested in marketing DES. The committee created the Master File, a collection of clinical evidence supporting claims about the safety of DES. Hines controlled the contents of the Master File by excluding all animal studies and including evidence only from short-term clinical studies. The historians Richard Gillam and Barton Bernstein argue that the Small Committee "thus effectively excluded unnerving evidence . . . based upon laboratory work with animals. As a result, a number of risks simply disappeared from sight."[34]

The drug companies hired as a lobbyist Carson Frailey, the executive vice president of the American Drug Manufacturers' Association. To generate evidence that DES was safe for women, Frailey worked with the drug companies to supply hundreds of doctors with samples to give to their female patients, thus creating a market for the drug even before its approval, as well as political pressure for that approval. In one interview Durrett wrote, "Dr. Nelson then took a bottle of 1 mgm Stilboestrol Tablets (Burroughs & Welcome [*sic*]) from his desk and asked for information concerning the preparation. He said Burroughs & Welcome has sent them to him and requested a clinical trial." These doctors treated thousands of patients with DES, and the doctors and their patients then wrote both to the FDA and to politicians (including President Roosevelt) asking them to speed the approval of DES.[35]

Carson Frailey persuaded fifty-four doctors from around the country to write to the FDA describing their clinical experiences with a total of more than five thousand patients. A Dr. Davis from North Carolina wrote to the FDA: "Is there a possibility that the manufacturers will be able to put this [drug] on the market any time soon? My experience with this drug has been so satisfactory that I feel it should be available." Only four of the fifty-four felt that DES should not be approved. Political pressure for DES approval increased, and Durrett could no longer persuade Campbell to be cautious with the drug. In August 1941, when it was clear that the FDA would approve DES, Shorr wrote to Durrett, "The last word in synthetic estrogens has not yet been said; it would be much nearer utterance if the industry had taken a more objective, scientific position as regards the need for further experimental development at that time."[36]

Despite the concerns expressed by many FDA medical staff, the FDA's drug chief, Theodore Klumpp, recommended that the FDA approve DES, and Campbell complied. Klumpp was a strong advocate of Campbell's partnership model, and he believed that the FDA should do its best to work with, rather than against, the industry, trusting manufacturers to make choices that protected consumers. Soon after pushing for the approval of DES, Klumpp went to work for Winthrop Laboratories, a major manufacturer of the drug — an early example of the government to industry "revolving door" problem that has so plagued the FDA.[37]

In 1940, FDA staff had used scientific uncertainty as a justification for refusing to approve DES, but that strategy was not proof against political pressure. The applicability of animal experiments to human safety was particularly contentious. As we saw in Chapter 2, when animal experiments on pesticide residues in the late 1930s had shown that those residues could be toxic, fruit growers had lobbied Congress to prevent the FDA from using results from animal studies to determine risks to people.[38] A federal court decision against the American Medical Association in 1938 made the FDA even more wary of engaging with drug companies over the applicability of animal models to humans. That year a company named Hiresta had developed a breast-enlarging estrogen cream called endocreme. Hiresta placed endocreme on the market shortly before the Food, Drug, and Cosmetic Act went into effect, so the company did have to apply for a New Drug Application or submit evidence of the cream's safety. The AMA, concerned about a possible increase in cancer risk from topical estrogen, published an editorial in April outlining the dangers of the cream, at which Hiresta sued the association for defamation. In court the FDA used animal studies to support the AMA's argument that estrogens were known carcinogens. The federal judge ruled against the association, arguing that animal studies failed to prove that estrogen cream would lead to cancer in women. Evidence of actual harm to specific women was lacking. This ruling against the AMA made other scientists unwilling to risk testifying against pharmaceutical companies. Dr. Robert T. Frank, for example, a scientist whom the FDA had asked to testify in a case against another topical estrogen cream, refused because of his bitterness over the way he had been treated when making his deposition in the endocreme case, including the "personal abuse" he had endured. The court case led the FDA in 1941 to abandon its planned campaign to regulate estrogen breast creams and made the agency wary of continuing to use animal studies in its case against the New Drug Applications for DES.[39]

After this court case, the FDA leadership decided to deal with scientific uncertainty through compromise: diethylstilbestrol would be available only via prescription — a novel idea at the time — and would be required to carry elaborate warnings about its toxic and carcinogenic risks. Through this regulation, designed to clarify the enforcement of the 1938 act, the FDA created a new class of drugs: those that were deemed dangerous enough to be sold only by prescription. While aiming to protect consumers, the regulation placed them, particularly menopausal women receiving estrogens, under the control of medical practitioners. As Watkins writes, under the new law doctors could "control not only the patient's treatment but also the degree to which she understood the nature and complexities of that treatment, since only the doctor was privy to the estrogen labeling information provided by the manufacturer."[40]

Initially, the FDA required that companies include detailed labeling information explicitly stating that DES was dangerous. One proposed Lilly circular warned that DES "should be used with extreme caution in women past 45 years of age in whom hysterectomy has not been performed, since continuation of menstruation in these women carries increased liability of the occurrence of uterine carcinoma." This warning essentially eliminated the entire market for the drug. Lilly staff complained to the FDA, "This is a very strong statement and if followed literally would almost exclude the use of the product." If all women with a uterus were excluded from the DES market, companies could hardly be expected to profit. The FDA backed down, allowing the warning to be limited to "patients who have a familial or personal history of mammary or genital malignancy."[41]

Because research had shown that DES could inhibit anterior pituitary function, which ultimately affects fertility, the FDA initially required companies to include a warning that women who wished to become pregnant in the future should not take the drug. On October 23, 1941, Merck asked the FDA to justify the exclusion of fertile women from DES use. The FDA commissioner wrote back that "first and primarily, we feel that promotional material regarding stilbestrol should be extremely conservative. . . . In our judgment, a few instances of sterility produced by this drug would be cause for serious concern on the part of all interested parties, and particularly so if the descriptive literature did not forthrightly inform the practitioner that such a consequence is a possibility." When Merck's staff objected that some doctors believed that DES would not impair fertility, Campbell replied, "A goodly number of qualified experts have recommended that the administration of the drug be restricted in the manner suggested by the sentence under discussion. A discussion as to the merits of the different viewpoints held by experts in this field would not appear to be particularly fruitful at this time."[42]

Merck tried to exploit the diversity of scientific opinions to advocate the position most beneficial to the company. Essentially, the industry position was that if a single researcher thought that a chemical might be safe, it should be allowed, even if a thousand other scientists thought the opposite. Capitalizing on scientific uncertainty became the key strategy deployed by lobbyists from numerous industries whenever they wished to defuse consumer concerns about risks from synthetic chemicals.

Cultural assumptions about women affected the development, approval, and marketing of DES. The urge to control the disorderly nature of the hormonal female body was closely linked to a sense that women were flawed by nature, unable to make rational decisions without the careful guidance of the experts. The FDA had hoped to find a compromise between industry pressure and scientists' concerns, limiting access to the drug through prescription and requiring that it carry a warning about DES risks. Yet these compromises foundered on the agency's assumption that women could not make appropriate choices.

Because the FDA staff did not trust female patients to evaluate medical information, regulators insisted that the warnings be made available on a separate circular that patients would not see. Doctors could get this warning circular only by writing to the drug companies and requesting it. Letters between companies and FDA regulators reveal that both groups feared that if women ever saw how many potential risks DES presented, they might refuse to take the drug. Worse, someone might take the drug, then sue the company, the prescribing doctors, and possibly the FDA if she developed cancer. One doctor expressed concern that label warnings "might bring repercussions on the physician should a patient develop carcinoma during estrogenic therapy."[43] Since few doctors requested the special circular before prescribing the heavily promoted drug, the distrust of female patients meant that few clinicians and fewer patients ever knew the full extent of the toxicity concerns regarding the synthetic estrogen.

One drug company tried to promote DES through a brochure titled "Reclaiming Years of Grace" that showed an older woman with hands stretched toward a young woman surrounded by little children in a tree-lined garden. The older woman stood alone, isolated in her cold business suit, yearning toward the vision of fertile youth. FDA staff told the company to revise the brochure, writing to the company director: "This Administration questions the propriety of the title 'Reclaiming Years of Grace' and the pictorial representation of women and children appearing on the front page of this circular." But while critical of such marketing approaches, FDA staff shared with the drug companies a wider set of cultural beliefs that defined women primarily as mothers and sexual partners, and therefore in need of medical treatment when these roles changed.[44]

The decision to treat menopausal symptoms with a potentially carcinogenic drug involved risk calculations that made implicit assumptions about gender. If a patient might otherwise die of a deadly disease, the decision to assume the risk of a toxic treatment rests on a calculation of relative costs and benefits that makes intuitive sense. A life-saving drug might have grave risks, but if the alternative is death, most people would choose to accept those risks. But what of menopause? Under what cultural assumptions do people assume that the risk of dying of cancer is worth taking to reclaim a woman's "years of grace?" FDA staff were certainly aware that risk calculations for DES differed from those for life-saving drugs. At times agency members objected to the assumptions about women made explicit in the marketing literature of drug companies. Yet they were unable to recognize that their own assumptions colored their risk assessments.

As the *Reader's Digest* article that opened this chapter illustrates, journalists viewed menopause as a crisis and a tragedy, a time of potential madness when women lost what was most essentially female about themselves: their reproductive potential. Physicians often shared these views. Many specialists saw women as creatures who were made irrational, almost childlike, by their hormones. One physician described his preferred treatment for the menopausal woman: "Her condition is carefully explained to her, her complete confidence and cooperation is sought. A simple sane hygienic regimen, mental as well as physical, is suggested. In addition, she is given adequate sedation, barbiturates or bromides, and

When the Ovary goes into Retirement

Wyeth's Estrogens, natural and synthetic, provide a convenient variety of precise dosage forms for estrogenic therapy:

WYETH'S Solution of Estrogens	WYETH'S Diethylstilbestrol (Stilbestrol)
Ampoules: 5,000 international units in 1 cc. corn oil 1 cc. ampoule—Boxes of 6, 50 and 100 5 cc. ampoule—Boxes of 1 each	*Tablets:* 0.1 mg., 0.25 mg., 0.5 mg., 1.0 mg. — Bottles of 40 and 500
Ampoules: 10,000 international units in 1 cc. corn oil 1 cc. ampoule—Boxes of 6, 50 and 100 5 cc. ampoule—Boxes of 1 each	*Ampoules:* 0.5 mg. in 1 cc. corn oil, 1.0 mg. in 1 cc. corn oil — Boxes of 6, 50 and 100
Ampoules: 20,000 international units in 1 cc. corn oil 1 cc. ampoule—Boxes of 6, 50 and 100	*Suppositories:* 0.1 mg., 0.5 mg. — Boxes of 12

Pharmaceuticals of John Wyeth & Brother, Division WYETH Incorporated, Philadelphia

"When the ovary goes into retirement": advertisement for DES
(American Journal of Obstetrics and Gynecology *46 [1943]: A-6)*

whatever other medication may be deemed necessary." Another, a Dr. Kohn, described one of his patients as having "gone crazy" when she learned that she had developed breast cancer after being treated with estrogens. Even after her cancer was gone, Kohn noted that this patient had "a great cancer phobia in spite of assurances that her cancer is cured; this she will not believe." Kohn added that the woman had been "receiving estrogens over a long period of time for premenstrual headaches and later on had received them for the menopause. There was no way to determine with any degree of accuracy the amounts she had received because so many people had been involved in the administration."[45]

This particular doctor dismissed the possibility that the estrogens the woman had taken over the years could have led to her breast cancer, arguing instead that she "had a carcinoma background since her mother had died of a malignancy." Thus it was the woman's fault for having a problematic genetic legacy, not the doctor's fault for the uncontrolled administration of estrogens to a woman with a family history of breast cancer. The doctor concluded by stating that "in spite of [the woman's] experience, he had no qualms about administering estrogens. . . . He then said he thought the Administration was going too far in the warning and caution direction."[46] Considering the estrogens these doctors gave to their patients, we might be tempted to conclude that it was the doctors, not the patients, who were irrational, willing as they were to use women's bodies for uncontrolled experiments.

Although Durrett and Campbell's cautious approach to the New Drug Applications shows they were concerned about the safety of DES for women as a group, FDA staff often treated individual women who questioned that safety with condescension. In March 1942, a woman wrote to the FDA describing painful symptoms that had developed after she started taking DES, asking why such "dangerous drugs should be permitted on the market." The FDA staff member who replied discounted her personal experience, assuring her that the drug had been "thoroughly tested in a number of clinics in this country before it was placed on the market and the results of the investigations demonstrated that the drug could be used with safety if it was used under the direct supervision of a physician."[47] The FDA, after its reluctance to approve DES, quickly adopted a public face of certainty, downplaying the continuing debates within the scientific and medical communities over estrogen safety.

CHAPTER 4

Bigger, Stronger Babies with Diethylstilbestrol

At the end of World War II, drug companies began exploring new uses for diethylstilbestrol. Pregnancy seemed to offer a vast potential market. Many Americans had expressed uneasiness about the wartime blurring of gender roles, and after the soldiers returned home, popular magazines and movies began extolling the virtues of femininity and domesticity. The FBI director, J. Edgar Hoover, urged women to leave work, marry early, and have children as a way of fighting "the twin enemies of freedom—crime and Communism": babies, not factories, were women's natural calling.[1] Ironically, this natural calling was to be achieved through medical technology, particularly the synthetic estrogen diethylstilbestrol, which was expected to help women have bigger, stronger babies.

Even with postwar improvements in prenatal care, miscarriages and premature births were frequent occurrences and a continued source of anguish for parents and doctors. Physicians had little sense of how to prevent such tragedies or to alleviate the grief parents felt at losing a pregnancy. When they were offered diethylstilbestrol as a potential solution in 1947, many within the medical community welcomed it with enthusiasm. Doctors, drug companies, and mothers all hoped to solve the problems of risky pregnancies with the new drug, which had the virtue of being a simple solution that was both easily marketed and highly profitable. After the Food and Drug Administration approved the use of DES for pregnancy, drug companies advertised it extensively, urging doctors to prescribe it "to make a normal pregnancy more normal." By 1957, advertisements in the *Journal of Obstetrics and Gynecology* were recommending DES for *all* pregnant women as a way of producing "bigger and stronger" babies.[2]

The postwar era had ushered in a tremendous boom in the development and marketing of synthetic chemicals, substances that we now know had the potential to cause fetal death and premature birth. In 2001, Matthew Longnecker and his colleagues at the National Institute of Environmental Health Sciences measured levels of dichlorodiphenyldichloroethylene, or DDE—a metabolite of DDT—in the stored blood sera of mothers who gave birth between 1959 and 1966, when DDT was heavily used in farming and DES use was rife. The greater the level of the estrogen disruptor DDE in the mother's blood, the higher the risk of preterm birth for the infant. One of the ironies of the DES story is that postwar chemicals acting as hormones may have been contributing to the problems that people hoped DES would solve.[3]

The FDA had initially refused to approve DES for menopausal women, not because scientists had any proof that the drug would harm women, but because they had no proof that it would *not* harm women. Within months, however, pressures on the FDA had forced the agency to reverse its decision. After World War II, when companies applied for approval to market DES for pregnant women, the agency once again had to balance the need for precaution against industry pressure. In 1941 the FDA had insisted that the drug was contraindicated for pregnant women because of possible risks to the uterus (not to the fetus itself) and that women who wished to have children in the future should never take DES. Why then did the FDA approve its use in pregnancy several years later? Why did researchers support it, why did the medical community prescribe it, and why did women agree to take it?

Political pressure on the FDA played a key role in the agency's change in policy. In 1941, FDA staffers had hoped that clearly defined limits on the use of DES would allow them to manage the substance safely. Within a year, however, the agency realized that enforcement of the initial limits on DES use was not going to be easy. One drug company even began selling DES over the counter in 1941, in direct violation of law, but the General Counsel's office decided not to prosecute, informing surprised FDA staffers that it wanted the first test case of the FDA's new regulatory authority to concern a drug that incontestably caused public harm. In a memo to all

Advertisement in a 1957 medical journal promoting the use of DES in all pregnancies
(American Journal of Obstetrics and Gynecology *73 [1957]: 14)*

stations, the FDA head office wrote: "While the possibility of developing a prosecution based on the over-the-counter sale of a dangerous drug received in interstate commerce is still being considered, only the most carefully chosen case will be discussed with the General Counsel's office initially. . . . When the courts are called upon to give judgment in a matter of this kind, we would like to have the decision based on a case which involves a more serious public health hazard than is exhibited by stilbestrol." The result was predictable. As drug companies learned that the government was unwilling or unable to enforce regulations against the new drug, they overwhelmed the agency with a more than a hundred New Drug Applications for various DES formulations and treatments. Pregnant women were soon to become a profitable market for DES.[4]

Research in the 1940s encouraged a hope that DES might help to prevent miscarriages. Elevated estrogen levels during pregnancy typically stimulate the hormone progesterone, which a uterus needs to sustain a pregnancy. Two Harvard physicians, Olive and George Smith, theorized that failures of pregnancy might be due to low levels of estrogen and thus treatable with DES. In 1943 the researchers Priscilla White and Hazel Hunt reported on sixteen diabetic women with a history of repeated miscarriages. After White and Hunt gave the women high doses of DES, most of their pregnancies went to term, prompting the researchers to declare the drug a success even though they had no control sample—women who had not been treated with DES—for comparison. A 1946 study by George Smith, Olive Smith, and David Hurwitz proposed a DES treatment schedule for diabetic pregnant women on the basis of a single woman's experience.[5]

Other researchers were dubious about the efficacy of DES in preventing miscarriages, and a lively debate arose in the medical literature about these experiments. White's work lacked a control group, and she did not indicate how many women took the hormone. The 1946 study provided data on neither efficacy nor safety, and critics pointed out that the research might not be applicable to other women. Yet given the prestige of Harvard and the influence of the *American Journal of Obstetrics and Gynecology,* many physicians accepted that the Smiths' research proved that DES was indeed a miracle drug for preventing "accidents of pregnancy."[6]

The willingness of doctors to believe that DES could prevent miscarriages came not simply from their longing to lessen the grief of pregnant

DES was promoted for a variety of conditions, not just menopause
(American Journal of Obstetrics and Gynecology *47 [1944]: A-3)*

women. The drug companies did their best to influence the way doctors perceived DES, manipulating scientific uncertainties and cultural beliefs in order to create a new market for their drug. Sales of DES for menopause were declining sharply as the new drug Premarin (an estrogen made from the urine of pregnant mares) took over the menopause market. Premarin was more expensive than DES but less likely to cause nausea and vomiting, so many women and doctors preferred it for menopausal symptoms. Hoping to create another significant market for DES, manufacturers turned to pregnant women.

To provide the FDA with short-term data on human effects, as well as to build a potential market for the drug, manufacturers sent DES samples to doctors, asking them to give the drug to their pregnant patients. In return, the physicians sent back case reports to the drug companies, who submitted the most favorable to the FDA.[7] A Houston physician named Karl John Karnaky became the most enthusiastic promoter of DES for use in pregnant women. He worked closely with drug companies, receiving their free samples. In return he sent them his positive results for submission to the FDA. As Karnaky later recalled, "The drug companies came to Houston . . . fed me and dined me . . . and I started using it." In 1946, Karnaky wrote to Squibb requesting more experimental drug samples: "I would like to have you continue sending me the 25 mg stilbestrol each month for at least 12 more months. . . . I would like to continue playing with stilbestrol and see what other uses we can work out for it. Personally, I believe it is a wonderful drug." In another instance, he insisted that DES was so safe that he would "finance the funeral costs 'up to $1000' of anyone who died from an excessive dose" of DES.[8]

Karnaky's research protocols were problematic, even by the more relaxed standards of the era. One of his reports described experiments done on fourteen healthy pregnant women in his privately financed clinic, experiments that included repeated X-rays and injections containing up to 24,000 milligrams of diethylstilbestrol. Karnaky described his treatment of one woman: "A normal patient, five months pregnant, was x-rayed and the uterus could be easily outlined on the x-ray film. Two hundred milligrams of diethylstilbestrol were given into the anterior wall of the cervix. Further x-rays were taken immediately following the injection and every 15 minutes afterwards." He noted that "all babies in the study were found to be entirely normal," even though he followed only five of the fourteen patients to term and did not report on the health of the other nine women or the results of their pregnancies. He did observe that all five babies "exhibited a darkening of the areolae around their nipples, labia, and linea albae, similar in intensity to that of their mothers, indicating that this effect of diethylstilbestrol . . . is shared by the fetus." Although drug companies later stated that no researcher suspected that DES could cross the placental barrier and affect the fetus, Karnaky's study, which the companies cited as evidence of the drug's safety, shows evidence of the opposite. McNeil Labs submitted a New Drug Application for DES on December 15, 1947, with safety claims based only on quotations from Karnaky's studies. The company's application insisted that "diethylstilbestrol is a non-toxic drug. It can be safely given in large doses for three to six months. . . . Remember, it is impossible to hurt anybody by giving large and continual doses of stilbestrol."[9]

Squibb sent twelve physicians (including Karnaky) free samples of DES to give to pregnant women, so that the drug company could present evidence of its safety to the FDA. Yet Squibb submitted records from only eight of the physicians. Although the eight physicians' records covered 108 pregnancies, doctors reported outcomes for only 16, ignoring the other 92 pregnancies. Of the 16 reported pregnancies, 9 resulted in full-term, healthy babies and 3 ended with premature births. Four babies were dead at birth. In other words, 43.75 percent of pregnancies treated with DES had adverse outcomes at birth. A full quarter of them ended in stillbirths. Squibb did not continue to monitor the surviving children, so we do not know what happened when they reached puberty.[10]

Many aspects of these studies are troubling. The company did not test the effects of DES against a control group to learn whether the drug actually lessened the risk of miscarriage. Using control groups was standard practice in clinical research by the late 1930s, but the FDA did not require them for New Drug Applications. Equally troubling was the insistence of researchers on seeing what they wanted to see. One patient was given an enormous dosage of DES, more than 69,000 milligrams over twenty-six weeks of her pregnancy. (In comparison, the usual dosage for a menopausal woman was 0.1 to 0.5 milligrams twice a week, usually for less than a year, totaling 10 to 50 milligrams.) The pregnant woman began bleeding during her eighth month and had a premature baby, which suggests that DES may not have worked for her. Nevertheless, without a control population, the doctor and Squibb both interpreted this result as a success, assuming that the baby would have died without DES.

Another doctor wrote to Squibb that one of his patients had also delivered a premature baby on DES treatment. Rather than interpreting this as evidence of DES's failure, the doctor instead blamed the premature birth on the women's personal behavior: she had gone shopping. The doctor's notes read: "P.S.: the fourth case delivered today. . . . She started in labor spontaneously delivering a premature baby of 5 lbs., which we felt was about 36 weeks gestation. She been taking 100 mg stilbestrol daily. The cause of the ruptured membrane, I am sure, was due to excessive shopping." Attributing a premature birth to "excessive shopping" rather than to the doctor's own experiments is disturbing enough, but what is even more surprising is that the FDA was willing to accept these results as evidence that DES was safe for fetuses.[11]

During the initial decisions about whether to approve DES for menopause, the FDA had wrestled with the results of hundreds of laboratory and clinical studies. The decision to approve the drug came only after a detailed engagement with conflicting scientific evidence. Given what we know today about DES, we might regret the decision, but we cannot deny that agency staff took seriously the evidence available at the time. But after World War II, when the FDA was trying to decide whether to approve DES for use during pregnancy, the agency's approach was less cautious. In early 1947 several drug companies filed supplemental New Drug Applications for the use of DES to prevent miscarriages. Gordon Granger, an FDA staff member who had been part of the deliberations about initial DES approval, was reluctant to approve it for pregnant women unless the mother was both diabetic and had a history of repeated miscarriages. Granger's concerns, however, rarely appear in FDA correspondence.[12]

During the decision to extend DES approval to pregnancy, agency staff memos contain no references to experimental research describing DES effects on animal fetuses. This is in striking contrast to the memos during the initial DES approval process, which reveal a near obsession with experimental animal findings. Studies in the late 1930s had shown that in certain animal species, DES was actually an abortifacient. In addition, when given to pregnant rats, mice, and chickens, DES had led to changes in sexual differentiation in their developing offspring. Many of these deformities had not been observable at birth but emerged only when the offspring reached the age of sexual maturity. Moreover, DES had increased the likelihood of reproductive cancers in those offspring when they reached sexual maturity. These studies had been well known to both the drug companies and the FDA, for they were among the ones whose abstracts were collated in the annotated bibliography the drug companies submitted to the FDA in 1940. These animal studies had played a significant role in the FDA's initial decision to reject the DES application for use in menopause. Yet they were completely ignored in the agency correspondence detailing the decision to allow DES during pregnancy. The New Drug Applications submitted by the drug companies fail to discuss them, and the agency documentation of the approval process does not mention the studies either.

Nor did the FDA or the drug companies consider the evidence of DES toxicity that had been published after the initial FDA approval. Between 1941 and 1946 endocrinologists and toxicologists published abundant evidence showing that DES had the ability to cross the placenta and cause abortions in laboratory animals, and potentially in humans as well.[13] None of it was mentioned by the drug companies or the FDA when, in 1947, the agency approved DES for pregnant women with diabetes. Almost immediately widespread use of DES for all pregnancies, not just pregnancies in diabetic women, began. Drug companies promoted the drug heavily, urging doctors to prescribe it even for healthy women "to make a normal pregnancy more normal." In the early 1950s clinical studies showed that DES failed to decrease the risk of miscarriages, but doctors continued to prescribe the drug for two more decades.[14]

Why did the FDA approve DES for pregnancy when no data showed that it was safe or effective? The historians Roberta Apfel and Susan Fisher suggest that the DES tragedy was an inevitable outcome of the state of medical research in the 1940s. In their view, doctors and researchers were well-meaning but did not understand the limitations of clinical research.[15] Yet the need for controls in clinical research was widely accepted by the early 1940s, and many medical researchers, toxicologists, and regulators were troubled by the research on DES. That their concern was not translated into policy was not simply because of their limited medical knowledge.

Other historians claim that the problem was that the FDA worked hand in hand with the industry, reneging on its regulatory responsibility.

As Gillam and Bernstein write, "Agency regulators, never doubting the drug's efficacy, clearly *wanted* to approve DES if the explosive safety issue could somehow be defused. By late 1940, perhaps responding to industry suggestions, FDA officials had hatched a plan to do precisely this. . . . Had the FDA wanted to avoid approving DES, it possessed the statutory authority to do so. Had the agency wished to resist the pressures for approval, it could always have delayed (it had done so before), rejected the Master File tactic (which was highly unusual), stressed the unsettling animal data, and given dissident experts a serious hearing. Put bluntly, FDA officials were predisposed, even eager, to approve the drug for human use, and such approval required a special effort—by no means inevitable—to accomplish this intended purpose." This argument ignores the intense scrutiny agency staff initially gave to the claims about DES and the skepticism many agency staff had about drug company claims. Although the FDA did accept the Master File, many agency staff and scientists, such as James Durrett, actively opposed the Small Committee's efforts to push the drug and continued to resist DES approval. [16]

But by the mid-1950s, the FDA was indeed becoming "weakened and demoralized" and was unable to stand up to industry pressure—one agent described the inspectors who led the FDA during the 1950s as "the rat-turd counters." The appointment of George Larrick as FDA commissioner in 1954 illustrates the increasing cooption of the FDA by the industry it was supposed to be regulating. Industry lobbying helped nominate Larrick, who according to the historian Philip Hilts, "believed in the mission of the drug industry, and had preached cooperation and harmony between the regulators and the companies as he rose through the ranks."[17] But the FDA had been much more skeptical of industry pressure during the debates over DES in the late 1930s and early 1940s. Political pressure played an important role in the outcome, but the FDA was never merely a pawn of the industry.

Other scholars believe that gender attitudes explain the approval of DES. The journalist Barbara Seaman writes: "Medical policy on estrogens has been to 'shoot first and apologize later'—to prescribe the drugs for a certain health problem and then see if there is a positive result. Over the years, hundreds of millions, possibly billions of women, have been lab animals in this unofficial trial. They were not volunteers. They were given no consent forms. And they were put at serious, often devastating risk." From this perspective, DES is a story about doctors and scientists experimenting on women as if they were little more than laboratory animals.[18]

Cultural beliefs about gender did shape DES approval and use. In an article titled "Synthetic Female Hormone Pills Considered Potential Danger" in the January 13, 1940, issue of *Science News Letter*—a publication aimed at the general public as well as researchers and physicians—the author admits the potential dangers of DES but then notes that "the new synthetic hormone has also been given with success to a few young women whom nature had partially cheated of their womanhood." According to this way of thinking, nature created one model of true womanhood, but it sometimes cheated individuals of their chance to be true women—and if it did, technology was the answer, risky as it might be.[19]

Similarly, the American studies scholar Julie Sze locates the reasons for acceptance of DES in gendered conceptions of the woman's role as a bearer of children, combined with what Sze calls "a utopian belief that technologies could harness and 'improve' on nature itself."[20] The doctor's comment that "excessive shopping" was the true cause of an adverse pregnancy outcome reveals that gender attitudes permeated hormone research and influenced FDA policy decisions. The national enthusiasm for procreation after World War II, combined with the frustration of the medical community over its powerlessness to combat miscarriages, encouraged many in the medical community to believe that a convenient regimen of pills could save babies.[21] But while gender attitudes help explain why physicians defined menopause and pregnancy as diseases in need of medical treatment, and those attitudes help us understand the policy decisions and some of their failures, gender attitudes alone cannot explain why the FDA staff discounted the mounting evidence that synthetic estrogens were causing fetal harm.

Changing conceptual models of fetal development influenced scientists and regulators, leading them to doubt laboratory evidence of DES toxicity. Scientists, doctors, and regulators were also guided by conceptual models of the placenta that made it difficult for them to imagine that a drug taken by a woman could cross the placenta and harm the fetus. The medical historian Ann Dally argues that many scientists and doctors assumed that the womb was inviolate, protected from the outside world by the placenta. Even though research in the 1940s revealed that estrogens *could* cross the placenta, most physicians continued to believe that it provided a barrier that protected the fetus from harm.[22]

One of the central conceptual puzzles scientists, doctors, and regulators faced in the 1940s and 1950s was imagining how something the fetus had been exposed to during development could lead to problems that were invisible at birth and emerged only in adulthood. Researchers had shown that exposing pregnant lab animals to DES could cause reproductive problems in the exposed fetus, but they could not comprehend such a thing happening in humans, for it violated their belief that the fetus was essentially determined by the genome and therefore invulnerable to environmental influences.

Although environmental concerns had played a central role in the development of embryology in the late nineteenth century, when investigators had tried to understand how the environment might shape embryonic development, by the early twentieth century a reductionist paradigm focusing on genetic influences had overtaken the ecological paradigm. The philosopher Jane Maienschein argues that "genetics brought a new form of preformationism. Instead of a dynamically acting organism taking its cues from the environmental conditions and from the way that cells interact with each cell division, the twentieth century brought a dominant and popular view that has often emphasized genes as programmed to carry the information of heredity, which was also the information necessary to construct an individual."[23]

Conceptual models alone, however, do not explain the synthesis of factors that emerged to create the disaster. Such models did not force the FDA to approve DES, and they did not force millions of women to take it. Cultural assumptions about the nature of women did not determine the FDA's response to the uncertain safety data, but such assumptions did frame the ways the regulators devised their policy compromises. Political and legal pressures on the FDA to approve the drug were great, and the agency's response to those pressures was constrained by confusion over changing conceptual models of toxicology and embryology, but it was not determined by this uncertainty. FDA staffers were initially skeptical about the drug companies' claims, but they based their skepticism on evidence that could not be defended in court and therefore could not stand up against political pressure. Cultural beliefs about gender and conceptual models of the body left the FDA vulnerable to political pressures from industry.

The importance of precaution when faced with uncertainty, and the difficulties of defending a precautionary principle under political pressure, are key lessons from the history of DES. In 1938, after a long battle, the FDA was given the regulatory power to require drug companies to show that their new drugs were safe. But no one knew what constituted "safe." No one knew what caused cancer or what effects a synthetic hormone derived from a carcinogen might have on people. No one knew how to translate studies on lab animals into a means of determining potential risks for adult humans, much less fetuses that one day might become adults. No one knew how women living in complex environments might respond to new drugs. Uncertainty wreathed every aspect of this brave new world of drug technologies and regulation. Yet none of these uncertainties made it inevitable that the FDA would side with the drug companies; that the agency did so emphasizes how fragile the illusion of safety provided by regulatory agencies is and how necessary it is for them to make precaution their guiding principle.

The approval of DES for pregnancy had another important repercussion. Karnaky's results on pregnant women helped persuade the FDA to approve DES in livestock feed, a decision that would soon expose nearly all people and wildlife in America to estrogenic residues.

CHAPTER 5

Modern Meat: Hormones in Livestock

In 1960 a Food and Drug Administration employee named Charles Durbin told a gathering of angry poultry producers, "Chemicals and drugs have revolutionized agriculture in the past 15 years. In animal husbandry growth-promoting chemicals permit the production of more meat with less feed; drugs . . . eliminate or control serious diseases. . . . Pesticides help the farmer control insects that would otherwise seriously affect his livestock. . . . Over 50% of the drugs used by the veterinarian and the feed mills weren't available to them in the early 1940's. Truly, agriculture has entered the chemical age." Durbin promised his audience that great developments were afoot in animal husbandry: "More hormones may find their way into tomorrow's feed; enzymes have been shown to improve the chick's ability to handle barley; tranquilizers have been reported to help birds during times of stress; we must take a new look at the appetite stimulators; and the Food and Drug Administration willing, we will see more and better medicinals being used in poultry feeds." Yet, he cautioned, his audience had to recognize that the modern "consumer is concerned with the question of chemicals in their food."[1]

Durbin understood that industry and government were struggling to control the narrative of modern drugs and chemicals in agriculture. As he told his audience, "We recognize that all constituents of food and feeds are chemical in nature, but unfortunately to many people the term 'chemical' denotes something poisonous or dangerous. It is up to all of us whether we be in Government, in industry, or the university to explain fully the story of drugs and chemicals in modern animal production."[2] This was a story that both the FDA and industry had begun to articulate decades earlier, but one that kept escaping their control. As disturbing new stories about the unpredictable risks of agricultural chemicals kept cropping up, consumers were not easily soothed.

Durbin's talk to poultry producers came just months after the FDA had banned chicken implants of DES under the Delaney Clause, a proviso within the 1958 Food Additives Amendment to the Food, Drug, and Cosmetic Act of 1938, which deemed no food additive safe "if it is found to induce cancer when ingested by man or animal." Poultry farmers were furious, but even with the ban consumers were rapidly losing their trust in the regulatory agencies. By 1960, what the journalist and rancher Orville Schell called "modern meat" was turning out to be not quite so nutritious or pure as envisioned.[3] Partnerships among the Food and Drug Administration, the U.S. Department of Agriculture (USDA), scientists, universities, industry, and farmers were unraveling as concern over chemicals erupted among consumers. The government had to negotiate growing consumer and scientific concerns over synthetic chemicals during an era when increased regulation received little political support and federal agencies were often overwhelmed by the businesses they were regulating.

The treatment of livestock with chemicals that Durbin described in 1960 had its roots in the agricultural industrialization that began in the 1920s. Believing that small farms were rarely profitable, engineers, agricultural economists, and government agencies had urged farmers to industrialize their operations through mechanization, standardization, and quantification. New chemicals could make standardization possible by reducing a farm's vulnerability to natural environmental fluctuations. As the historian Deborah Fitzgerald shows in *Every Farm a Factory*, a complex set of relationships developed among the USDA, agricultural equipment manufacturers, university experts who envisioned a model farming landscape in which farms were run like factories, and the farmers who actually tried to effect much of this transformation, often at significant cost to their families and their farms.[4]

World War II had been a boom time in the development of chemical technologies for military purposes, and these quickly expanded to civilian uses. Animal husbandry was no exception. After the war, the new technologies were rapidly appropriated for agriculture, boosted by the enormous growth in advertising during the postwar era. The expansion of wartime technologies into postwar agriculture was shaped by what James C. Scott terms a high-modernist ideology, "a strong, one might even say muscle-bound, version of the self-confidence about scientific and technical progress, the expansion of production, the growing satisfaction of human needs, the mastery of nature (including human nature), and above all, the rational design of social order commensurate with the scientific understanding of natural laws."[5]

Consumer marketing promoted technological optimism, particularly for the chemical miracle promised by advertisers, doctors, and extension agents. This optimism existed in an uneasy relation with Cold War anxieties about threats to the free-market system, risks stemming from new chemical technologies, and the changing role of women in American society. The use of synthetic hormones in livestock exploded in part as a response to these anxieties, yet synthetic chemicals also served to increase them.

As the social historian Amy Bentley argues, meat held powerful symbolic meaning for Americans, particularly during wartime. American meat was seen in gendered terms: meat made men manly, strong, and able to fight. A navy pamphlet extolled its virtues: "Meat is one of the major providers of life — essential complete proteins needed to repair body tissue as it wears out, and to furnish the building blocks for new muscle and sinew in husky, hardy men who follow the sea." Without meat, the folks at home might lose their will to stand firm. Bentley writes that "being deprived [of meat] meant much more than going without meat for dinner; the absence questioned the very health and strength of American society."[6]

During World War II, pharmaceutical companies requested that the FDA approve the use of diethylstilbestrol to treat certain veterinary conditions in livestock, and the agency agreed, in part because of the perceived urgency of producing meat for fighting men. Yet because of concern about the potential risks to soldiers who might consume estrogen residues, the FDA explicitly forbade treatment of livestock that might be eaten. Could eating estrogen-treated meat turn men into girls? No one wanted to find out during a war.

When companies tried to push against wartime FDA restrictions on the use of synthetic hormones in livestock, the FDA initially insisted on precaution, arguing that the absence of evidence of harm did not prove safety. In 1944, for instance, one company wrote to the FDA asking about extending veterinary use of DES to include chemical castration of roosters intended for human food. The FDA refused, pointing out that safety studies had not yet been done and the company would have to submit a full New Drug Application with tests showing that any remaining chemical residues would be safe for consumers.[7]

In 1945 the feed company Wick and Fry requested permission to manufacture "sex hormone pellets" to be inserted into chicken necks. Rather than submitting the required safety tests, the company offered theoretical arguments purporting to demonstrate why residues could not harm consumers. According to Wick and Fry, the pellets posed no danger because modern housewives would surely throw the chicken necks out, thus avoiding eating the pellets as well. The FDA was unimpressed with the company's logic, responding: "We have no way of knowing that the poultry man will do as your label directs. He may find it easier and quicker to implant the pellets a little lower in the neck. Second, the housewife, who has no way of knowing that her purchase may have a pellet in its neck, may be a frugal person who cuts off the head right behind the skull and uses the entire neck meat in the gravy. . . . In view of these comments and the fact that small quantities of diethylstilbestrol may produce very undesirable effects in humans, you can appreciate that the use of the drug for the purpose of fattening or tenderizing poultry causes us real concern." The FDA insisted on empirical data rather than theoretical arguments about what model housewives and model farmers might do in an ideal world.[8]

As the war came to an end, pressure began to build on the FDA to approve use of diethylstilbestrol in animals intended for human consumption. The wartime meat rationing had ended in the United States, but food shortages throughout Europe threatened to lead to famine, which many people were afraid might destroy the peace. Grains were being used to feed livestock rather than people, threatening shortages, but government officials worried that Americans would be unwilling to voluntarily reduce their meat consumption in order to make more grain available for human food. Rather than reinstitute rationing, the government encouraged research partnerships devoted to learning how to increase meat production while retaining enough grain to prevent famine. The answer appeared to be hormones, which promised more efficient feed utilization.

Because animals treated with estrogen fattened up more quickly on less grain, science might allow Americans to eat more meat without guilt.[9]

Starting soon after World War II, professors from agricultural research stations, feed mills, and agricultural products companies wrote numerous letters to the FDA inquiring about diethylstilbestrol use in livestock. Poultry was the first target, for roosters and turkeys responded readily to DES implants.[10] In February 1946, internal memos within the FDA showed that staff remained skeptical about the use of diethylstilbestrol in poultry. When Wick and Fry insisted again that pellets were safe because they would be implanted in chicken necks, which middle-class housewives discarded, one staffer scribbled on a memo: "Some people do use the heads of poultry for food!" Another memo in the same file expressed doubt that any kind of residue testing would be accurate enough to show whether the meat was safe.[11]

Throughout 1946, the FDA rejected New Drug Applications submitted for poultry, stating that "no information has been offered to show the amount of diethylstilbestrol remaining in the tissues of treated birds. Until it can be clearly shown that no significant quantity of the drug remains in the tissues which might be capable of producing undesirable effects in human consumers, we will not be disposed to consider any application for the diethylstilbestrol use with this purpose." FDA staff insisted that determining accurate residue levels was the responsibility of the manufacturer, not the problem of the government.[12]

In January 1947 the agency reversed course and agreed to allow diethylstilbestrol to be used in poultry implants. None of the problems discussed in the correspondence from the previous several years had yet been fixed. The only research that drug companies offered in support of DES pellet safety actually showed that estrogen residues did migrate from the pellets into meat intended for human consumption. The New Drug Application claimed that "the amounts of the synthetic estrogens deposited in the tissues were insignificant from the standpoint of human consumption," yet the company presented no evidence in favor of this assertion.[13]

Why did the agency suddenly allow DES implants in livestock, when regulators had resisted for years? Pressures to increase meat production after the war were certainly great, but concerns about the risk of estrogens for men had initially led the FDA to resist these pressures. By 1947, however, DES began to seem much safer to the FDA. Because Dr. Karnaky and

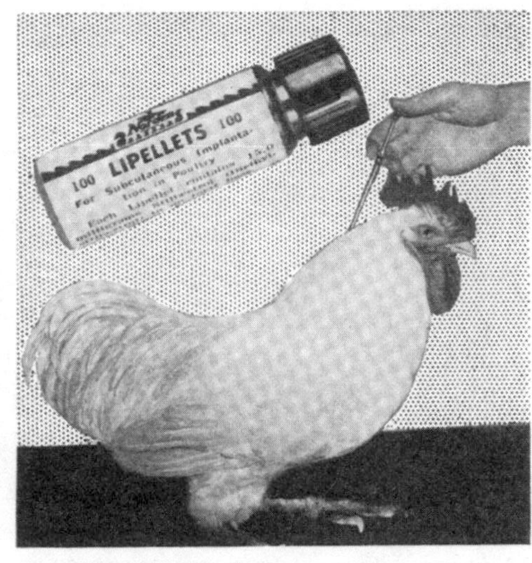

Increase the Value of Cockerels, Roosters, Old Hens and Turkeys *by* "Lipellizing"

with *Norden*

LIPELLETS

(Each LIPELLET contains 15 mgms. Stilbestrol)

Increases muscular and subcutaneous fat

Tenderizes old birds

Feminizes cockerels

Implant subcutaneously 3-5 weeks before marketing

100 . . . . . . $1.75 1,000 . . . . . . $15.75

Pellet Injector $1.75

Ask for Free Booklet

*Advertisement for DES use in poultry as a way to tenderize old birds and feminize cockerels (*Veterinary Medicine *42 [1947]: iv)*

other researchers had treated pregnant women with large doses of DES and no deaths had yet resulted, FDA staff began to argue that small doses presented little risk.[14]

Not all scientists or regulators agreed. Immediately after the FDA approved the chicken implants, Canadian regulators wrote to the federal government, urging that the FDA be extremely careful with the use of diethylstilbestrol in animals. A staff member from the Canadian Department of National Health and Welfare wrote, "We have been working on the problem with the poultry division of the Department of Agriculture and our results show that there is a residue of the estrogen in the cockerels, sufficient to change the vaginal smear of the menopausal woman. Of course this is not evidence of any harmful effects but it is possibly an undesirable reaction for some people. . . . We were planning to publish these results and are wondering if any of the results from your division had been published and we had overlooked them." The Canadians, that is, had data showing that a synthetic estrogen implanted in chicken necks was so powerful that residues left in the meat could change the vaginal smears of the women who ate that meat.[15]

Why were Canadian regulators able to find evidence of residues and their effects, while U.S. regulators could not? In part, it was because Canadian laws and U.S laws were different. Canadian regulators conducted their own research to determine the safety of drugs and synthetic chemicals. In contrast, U.S. regulators were required to use research performed by the companies and reported by the companies as well. Results not favorable to the companies might never be reported to the government. Research design was also vulnerable to the desire of the companies not to find adverse results. The Canadian government tests found estrogenic effects of residues, whereas the companies in the United States did not test for estrogenic effects, yet they continued to insist that no such effects existed. The FDA attempted to defend its position by finding flaws in the Canadian research, suggesting that if Canadians were observing estrogenic effects, it was because their research protocols were incorrect, not because American tests were insufficient.[16]

In July 1947, Dr. J. S. Glover of the Canadian government responded to the FDA, noting that Canadian research "indicates that the feeding of tissues from treated birds has repeatedly produced estrous changes in clinical tests with aged women. As we receive numerous inquiries regarding the efficacy and safety of so-called 'chemical caponizing' I should appreciate it very much if you would clarify the issue for me."[17] The FDA reply noted that "the periodical you refer to is not immediately available to us," and went on to assure Glover that "according to our medical staff a 15 mg pellet of this drug when used as directed would not cause any particular harm in humans from consuming such treated birds." Glover was not reassured. He replied that "a technical paper dealing with the subject of the administration of female hormones is shortly to appear in *Endocrinology*." This was the major journal in the field, one that would certainly be "immediately available" to the FDA.[18]

After the October 1947 issue of *Endocrinology* appeared, the FDA did move to forbid the use of DES, but only in chicken feed; farmers could continue to implant DES pellets in the chickens.[19] The Canadians had tested DES only in feed and had not investigated whether DES in pellets produced the same response. Rather than insisting that pellets be tested, the FDA accepted industry assurances that DES from pellets would not contaminate the meat, even if DES in feed did. For years the FDA continued to insist on something that made little scientific sense: that although diethylstilbestrol in feed would accumulate in the poultry's fatty tissues and pose a danger to humans, DES administered in pellets would not.

When challenged, C. W. Crawford, associate commissioner of the FDA, argued that "it is possible to exercise a rigid control over the dosage in the [pellet] process and under these circumstances the estrogen does not accumulate in those portions of the treated bird which are consumed by human beings."[20] A number of assumptions about the possibilities of scientific control are embedded in Crawford's statement. First, Crawford assumes that technology can offer enough control to sidestep dangers posed by pollutants. He also assumes that farmers will always follow instructions exactly, that consumers will always eat what they are supposed to eat, and that companies will make perfect pellets that will always release an exact dosage. None of these assumptions was based on empirical evidence. The FDA had never received or examined data showing that pellets did release a reliable and controllable dosage or that this dosage did not accumulate in tissues, even if it were controllable and reliable.

The use of DES in food stirred up a host of anxieties about threats to prevailing gender norms. Journalists tried to reassure consumers that the hormone would not demasculinize men the way it did roosters. One popular article noted in February 1948 that when a rooster was fed DES, he was soon transformed into a "fat, complacent fowl that would rather cluck than crow and [would] ignore the prettiest hen that passes by. . . . Roosters forget about crowing and prefer to cluck like matronly hens. They have little interest in breeding and become content to sit quietly and get fat. . . . Tests, so far, show that the drug does not affect humans who eat the treated birds. . . . Thus the housewife need not fear that if her husband eats a stilbestrol chicken he will give up golf and hunting and start knitting sweaters."[21] The journalist's "so far" suggested that men who ate too much treated meat might indeed be wary of finding themselves a little less manly, more interested in domesticity than in dead animals. One of the great ironies of promoting DES for livestock was clear: meat went hand in hand with manliness, but if a female hormone were used to produce more meat, what happened then?

After approving DES for use in chickens in 1947, the FDA soon began to receive questions about its effects on men's reproductive health. The drug appeared to be causing problems among chemical plant workers, farmworkers, restaurant workers, and consumers. In 1947, Arapaho Chemicals of Colorado wrote to the FDA: "Our Company has recently been approached in regard to manufacturing stilboestrol . . . as raw materials for pharmaceutical formulation. We know that these materials are all readily absorbed through the skin and by inhalation. It is our belief that the physiological effect of these materials would constitute a decided industrial hazard. In order to properly evaluate the advantages of undertaking the manufacture of synthetic estrogens, it is necessary that we obtain as much information as possible about them in regard to the seriousness of the health hazard involved, recommended precautions for handling, treatment of affected individuals, cumulative effects, etc. We are particularly concerned over the possibility of carcinogenesis through long continued contact with stilboestrol."[22]

The FDA responded by suggesting that the company hire old men, presuming that they would not mind being devirilized by their jobs: "It is our understanding that excessive exposure to the substances may cause marked disturbances of the menstrual function in women and have a devirilizing effect in men. For this reason it might be feasible for you to consider the employment of old rather than young men."[23] At the same time, then, that the FDA was reassuring the media and consumers that livestock treated with DES was safe for human consumption, the agency was acknowledging that male workers might become infertile and grow breasts, while female workers might have their menstrual cycles disturbed. In a handwritten note on the letter from Arapahoe Chemicals, an FDA staff member mentioned, "I've had two previous inquiries along this line, the most recent being from Ortho products. The greatest complaint concerns women who have deranged menstrual cycles, and excessive bleeding. . . . There is no question as to the de-virilizing effects on males, and it may result in permanently lowered virility and sterility." Yet rather than reconsider approval of a drug that "deranged" menstrual cycles, FDA staffers began to argue that people who complained of sterilization and cancer were failing to provide "material proof" of their case, even though legally the burden of proof was on the manufacturer. Another FDA staff member added on the letter: "The data on carcinogenesis is meager and many published opinions are not properly backed up on the facts. I personally doubt if most or not all of the people who have raised the question have failed to provide material proof of their contention. The other side, however, has adequate proof of the lack of carcinogenic activity of the estrogen."[24] These notes suggests that the FDA was beginning to view absence of proof of harm as a form of proof of safety, an interpretation that Harvey Washington Wiley had long warned against.

The FDA staff continued to assure consumers, doctors, and scientists that the synthetic estrogens had been shown to be completely safe. Just ten days after these internal memos circulated in July 1947, a professor at Cornell wrote to the FDA asking whether diethylstilbestrol was really safe in livestock. In response, the agency did not mention the Canadian research or the concerns voiced by chemical manufacturers. It simply dismissed the entire question of residues by referring to Karnaky's work and claiming that if the amount of DES in the original pellet were safe for a human, then the residues must also be safe.[25] Karnaky had long been an enthusiastic promoter of DES use in women, and his reductionist logic was extended to livestock implants: if huge doses did not kill pregnant women, small doses must be safe for everyone. The FDA moved from an insistence on zero residues to a refusal even to require measurement of the residues or to consider that residues might be problematic.

Troubling findings began accumulating. A male restaurant worker in

New York grew breasts after eating the heads of chickens implanted with DES pellets, and his case became immortalized in a medical textbook. Mink ranchers began complaining to the FDA that their mink were made sterile by residues from the necks of the implanted chickens. The FDA discounted these complaints, testifying that "a few mink ranchers have alleged that their breeding animals were rendered sterile after having been fed the discarded heads of poultry which were implanted with diethylstilbestrol pellets. As yet we have seen no satisfactory data of a factual or scientifically acceptable nature showing that the offal from birds implanted with these pellets will actually cause sterility in minks or any other animals."[26] Yet mink ranchers could not be expected to produce scientific data; they were ranchers, not scientists.

Finally, five months after repeated reports of problems, in 1950 the FDA began checking chickens. After finding numerous cases where residue levels violated the law, agency staff seized more than 22,000 kilograms of chickens containing high levels of DES residues. Some birds had as many as four pellets in a single neck. The FDA had vigorously denied that chicken meat could contain residues; when mink ranchers had sent their informal evidence, the FDA had simply denied its validity. Yet when FDA staff members finally went out and collected their own data, they found that the scientific models of what ought to be happening were not supported by empirical evidence.[27]

After residues were discovered, consumer pressure forced the FDA to begin restricting DES use in poultry, although nine years passed before the FDA finally called for a "voluntary discontinuance" of the hormone. The use of DES in chickens was eventually banned in 1959, but the poultry industry took the FDA to court, and the ruling was not upheld until 1966.[28] Even as the FDA began the long process of banning DES in poultry, the agency approved its use in cattle. What rationale made residues from poultry dangerous and residues from beef a reasonable risk for consumers to bear? The answer lay in the difference between synthetic chemicals delivered by implanted pellets and those delivered by feed. These differences might seem technical—as indeed they were. Even in the face of growing evidence to the contrary, the agency's staff continued to believe that precise technological control over hormone residues would be possible. Yet ecological complexity, natural variability, and simple human error meant that such control remained an illusion.

In *Cancer from Beef,* the agricultural historian Alan Marcus examines the growth and development of DES use in beef. Briefly, in 1947 a graduate student at Purdue named W. E. Dinusson experimented with DES implants in heifers. He had noted that spayed heifers put on weight more slowly than intact heifers, so he hypothesized that if having less estrogen than normal slowed growth rates, having more estrogen than normal might increase them. He found that DES did lead to faster growth in young cows, but the hormonal side effects such as vulvar swelling and mammary development were too great to recommend use. Implants of DES "produced a nymphomaniacal stance" in heifers, and "the meat and liver of these slaughtered experimental animals retained significant estrogenic activity, a factor that rendered safe human consumption problematic."[29] DES pellets seemed like a dead end for cattle.

Several years later, three researchers at Iowa State (W.H. Hale, his graduate student C. D. Story, and Wise Burroughs) decided to see whether DES could be used in feed, rather than in implants. DES supplementation in feed had not worked for chickens because their livers quickly broke it down and because their food consumption could not be closely controlled. Cows and sheep, on the other hand, are ruminants, metabolizing nutrients in unique ways. Australian researchers had long noted that certain clovers with high levels of phytoestrogens led to striking changes in the ruminants that grazed on them. Some sheep had repeated miscarriages, which suggested that plant estrogens might be harmful to fertility. But other sheep and cattle had higher growth rates when eating estrogenic clovers, and thus supplementing cattle feed with estrogens might have interesting possibilities, particularly in animals headed for the butcher whose future fertility was therefore of little concern. According to the agricultural hormone researchers A. P. Raun and R. L. Preston, Hale noticed a report in a British journal that indicated that oral DES was rapidly "detoxified" in chickens but not in cattle; thus feeding DES to cattle might be useful even though it was useless in chickens. Hale and Story decided to conduct experiments with DES added to cattle feed. In one study, they found that lower levels of DES in feed improved weight gain, but higher levels had no effect. "The responses in the first two studies are unexplainable," complained one critic, because they violated normal dose-response theories. Yet given what we now know about the effects of endocrine disruptors, the results were not surprising.[30]

In 1953, Burroughs published a report showing that "cattle gains could be increased substantially and that feed costs could be reduced materially by placing 5 mg or more of DES in the daily supplemental feed fed to each steer." Burroughs concluded that DES feeding led to 35 percent increases in growth and a decreased feed cost of 20 percent—astonishing results if they could be reproduced.[31] Burroughs's university, Iowa State, was intrigued by the financial possibilities of this research. The university entered into complex patent negotiations with the drug company Lilly and did its best to create a market among farmers for the hormone. Burroughs announced his discovery at a special Iowa Cattle Feeder's Day, and Iowa State skillfully managed the publicity. As one witness to the announcement recalls, "Publicity about a new discovery resulted in a huge and unexpected crowd (over 1,000)."[32]

Lilly's New Drug Application for DES in beef feed was about to come up for approval under a new FDA commissioner, George Larrick. Larrick saw himself as an advocate for business, and industry had pushed hard for his appointment. Winton B. Rankin, who served as Larrick's deputy commissioner, recalled that Larrick had won the post "because the drug industry came to his support. . . . He had a very warm spot in his heart for the responsible members of the drug industry." Theodore Klumpp, the former FDA official who had become a pharmaceutical company director after pushing the DES approval through, commented that during Larrick's tenure (which lasted until 1965), "the regulated industries had felt their relationship with FDA was beginning to approach what it should be in a free society and in accord with the best traditions of the Republic."[33]

Under Larrick's guidance, the FDA granted approval for DES in cattle feed on November 1, 1954. Just a year had passed since the initial report from the feeding studies. To protect consumers from residues, the FDA required only that farmers withdraw the drug from the cattle feed forty-eight hours before slaughter. No significant DES residues, the FDA believed, could remain in the meat after the forty-eight-hour withdrawal period, assuming that cattle quickly excreted feed additives in urine and feces. Lilly worked with Burroughs to devise tests for the FDA to show that measurable residues would not remain after forty-eight hours. Yet these tests were not particularly sensitive, for they could not measure DES levels in beef that were high enough to cause biological changes in tissues, even though it was possible to measure those levels in poultry. Neverthe-

When fed to steers on high-corn rations, *Stilbosol* boosts gains as much as 37% while slashing cost of gain as much as 20%. *Stilbosol* is truly called the *new* beef profit builder.

SELECT SUPPLEMENT IN SAME WAY

You still select the type of supplement you'd normally pick to do your particular job . . . whether it be complete, high protein, lower protein, or whatever. Just be sure that *Stilbosol* is added. Look on the bag or ingredient tag.

Stilbosol will make both a good and a poor cattle supplement more efficient. However, *Stilbosol* will never make a poor supplement the equal of a top-quality supplement that contains *Stilbosol*. *Stilbosol* merely piles its benefits *on top of* those originally built into the beef supplement.

Stilbosol is not a substitute for such nutrients as proteins, minerals, and vitamins. Rather, *Stilbosol* appears to improve the nutritive properties of any and all supplements and feeds consumed by beef cattle fed for the market.

ONLY FOR BEEF CATTLE FATTENING RATIONS

And, here's a very important point. *Stilbosol*-fortified rations are not now designed or recommended for any kind of livestock except beef cattle to be fattened for slaughter.

For the present, that rules out its use in rations for dairy cattle, beef breeding stock, sheep, swine, poultry . . . or anything else.

The sale of *Stilbosol* and its use in beef fattening supplements are subject to Federal Food and Drug regulations.

Feeding of the final *Stilbosol*-fortified supplement should be done with equal care. Follow the manufacturer's instructions. *Insist on beef fattening rations fortified with* Stilbosol.

Stilbosol

makes the difference

SO . . . BE SURE—Look for *Stilbosol* (Diethylstilbestrol Premix, Lilly) on the bag or ingredient tag of the beef fattening supplement you buy.

Stilbosol

(Diethylstilbestrol Premix, Lilly)

Stilbosol is Eli Lilly and Company's trademark for Diethylstilbestrol Premix, Lilly. *Stilbosol* is compounded under license from the Iowa State College Research Foundation, Inc.

ELI LILLY AND COMPANY, AGRICULTURAL PRODUCTS DIVISION, INDIANAPOLIS 6, INDIANA

*"Stilbosol makes the difference" for beef profits (*Flour & Feed *55 [1954]: 15)*

less, the FDA was satisfied that, at least under ideal laboratory conditions, beef would be safe from DES residues if the withdrawal periods were adhered to, despite the fact that in the case of poultry farmers it had turned out that few farm operations were able to replicate laboratory conditions in which an exact withdrawal period led to pure meat. Most chickens treated with DES had been contaminated, either because of simple negligence or outright duplicity. But instead of learning from these chaotic results that complete technological control over hormones was not possible in the real world of agricultural production and consumption, the FDA insisted that the problem lay with farmers, not with the hormone itself.[34]

A month later, DES went on the cattle-feed market as Stilbosol. Lilly and other manufacturers marketed DES feeds widely to extension agents, farmers, and farm publications. As one observer noted, "Cattlemen turned to the enhanced feeds in droves." By late 1955, less than a year after DES had been approved, fully half the cattle in America were receiving DES. Soon, 80 to 95 percent of cattle received DES.[35]

The research, FDA approval, and marketing of DES did not happen by accident; they emerged as part of a complex partnership linking drug companies, universities, and federal agencies. Researchers at Purdue had initiated studies on DES in cattle, but the Purdue administration did not believe that commercialization of new technologies was part of the university's academic role. Iowa State, however, was willing to work with drug companies to profit from academic discoveries. After a series of confidential meetings, in 1954 the university decided to grant an exclusive license to Lilly under a university patent. Because of these partnerships, DES transformed beef-production practices. As Marcus writes: "The case of DES seemed to be a model of the application of the partnership idea. A college scientist uncovered a new technique, pharmaceutical scientists produced the drug, feed-manufacturing scientists compounded the material as a premix, federal scientists approved its use, agricultural college scientists publicized it by demonstrating its utility, and farmers made use of it. That type of expert-based interaction had been the model for 'progress' since the 1920s. With respect to stilbestrol, little in the mid-1950s seemed to undercut faith in that model." The synthetic hormone was a critical factor in the rapid expansion of feedlots, according to Marcus. Beef consumption nearly doubled during the drug's heyday between 1954 and

*Advertisement for DES in cattle feed (*The Cattleman *42 [1955]: 38)*

1972. Yet even as consumers ate more meat, they also worried about what might be lurking in that meat.[36]

During the 1950s and early 1960s, the FDA came under increasing pressure from consumers concerned about chemical residues in their food, particularly pesticides and hormones. In 1950 workers in the agency's Division of Pharmacology found that DDT was showing up in the fat of ordinary Americans. Consumers wanted to know where it was coming from. Baby food manufacturers began testing their sources and announced that they could not find vegetables free of pesticide residues. When DDT residues were found in milk—the epitome of a wholesome food for children—consumer concern about synthetic chemicals in livestock production began to build.[37]

Meanwhile, the regulatory agencies were confronted by an increasingly industrialized food system, whose producers were transforming U.S. agriculture into centralized agribusiness. Agencies wrestled with the question of how to deal with conflicting demands from consumers and producers. No one seemed certain whether the primary mission of the regulatory agencies was to protect consumers by controlling industry—or to promote industry by serving as its partner. New technologies, the federal agencies hoped, would allow them to fulfill both missions.

In a 1951 speech, an FDA representative, Dr. John H. Collins, spoke to the livestock industry, warning of growing consumer concern: "Increased public interest in what goes into foods is shown by the recent activities of the House of Representatives' Select Committee to Investigate the Use of Chemicals in Food Products. But at the same time there has been a growing tendency among livestock people to employ drugs to promote fattening, stimulate milk production, and bring about other physiological changes in domesticated animals and poultry."[38] As Collins noted, increasing concern and increasing chemical use were about to collide. Regulatory agencies were caught between political pressure from lobbyists and growing pressure from consumers. To the FDA and the livestock industry, the answer seemed to be in closer partnerships between industry and universities, with the federal agencies helping to foster scientific research and promote new drugs. In 1954 this faith in science as a way of defusing consumer concern about pesticides found expression in the Miller Pesticide Amendment, which allowed residues of toxic chemicals to be present on food as long as they were below a tolerance level assumed to be safe. The FDA was responsible for setting tolerances, which were defined as the levels of residues that people could, in theory, tolerate without harm.[39] The amendment reflected a belief that scientists could measure risk and that something that could not be measured was not dangerous.

But as postwar concerns about the chemical age grew, fed by anxieties over radiation exposure, nuclear technology, pesticides, and new drugs, more researchers begin joining with consumers to express doubts that *any* residue of a known carcinogen could be safe. In 1955 a report in *Science* warned about mouse feed that had been prepared in a mill that had previously been used to prepare cattle supplements. The mouse feed had thus been contaminated by DES, which had led to serious reproductive problems in the mice. This finding stirred renewed fears about hormone residues in food.[40]

At a 1956 FDA symposium on medicated feeds, several scientists argued against giving DES to livestock, citing decades of experimental evidence on laboratory animals that raised concerns about their effects on humans. In particular, scientists were concerned about experiments showing that small doses of DES could be more effective at inducing cancer than large doses, just as lower doses of DES were more effective at inducing weight gain in the cattle. This had troubling implications for animal feed. Arguments for the safety of exposures to small amounts of DES through feed relied on the assumption that since pregnant women did not seem to be harmed by the large doses of DES given to them, much smaller doses must be even safer. If DES and other hormonally active synthetic chemicals did not follow normal dose-response relations, the scientific basis of their regulation was undermined.

The cancer researcher William E. Smith opened the 1956 symposium with a joint paper titled "Possible Cancer Hazard Presented by Feeding Diethylstilbestrol to Cattle" urging the government to reconsider its approval of DES for cattle. Using animal studies as the basis for his argument, Smith framed a precautionary principle that would exclude the use of DES in any animals intended for human consumption. Smith dismissed the FDA's argument that because no residues could be measured in beef, there was no cause for concern. The mechanisms of cancer initiation were only beginning to be understood in the 1950s, but Smith reminded his listeners that even if they could not measure DES residues in the meat of treated animals, the hormone could have catalyzed changes in the cells that would eventually lead to cancer in the consumer.[41]

Smith noted several key ways that DES in the food supply could be problematic. First, DES appeared to bypass the body's ability to break down toxins and thus to heal itself. Viruses in mice could cause tumors, and though normal mice could fight those viruses, DES-treated mice were unable to resist them. Understanding cancer and environmental exposure could not be a matter simply of examining correlations between cancer and current residues because complex indirect effects on the body's own healing mechanisms were difficult to detect.[42]

Smith also questioned the FDA's reassurance that the DES was safe because the doses were low. Smith pointed out that meat from a steer fed the recommended amount of DES would still contain significant residues. Although the residues were lower than could be detected by standard testing, they were fourteen times greater than the amount of DES that had induced cancer in mice. Smith acknowledged the core argument made by the FDA and livestock interests: normal DES residues were only a fraction of the body's own natural estrogen levels. Yet mice treated with such low levels of DES had indeed developed cancer. Low did not mean safe.[43]

Industry and FDA staff had minimized the potential risks of low-dose exposure to DES because pregnant women had appeared to tolerate higher doses. If a lot of DES was safe, wasn't a little DES even safer? Smith refuted this logic, pointing out that the fact that DES was *not* dose dependent in its effects suggested that standard toxicological paradigms were inadequate for regulating its safety. Smith was particularly concerned about research showing that chronic, low-dose exposure to estrogens had induced cancer in mice even though intermittent high doses had not. Putting DES in the food supply, Smith pointed out, had resulted in precisely this continuing, chronic, low-dose exposure in humans that had caused cancer in mice.[44]

Smith was particularly concerned by how quickly DES had permeated the food supply of the nation. In 1956 more than thirty million chickens per year were implanted with DES pellets, and over half the feedlot cattle received DES. In his view, this amounted to an uncontrolled experiment that the government was allowing to be conducted on the nation's citizens, with a drug known to induce cancer.

The person chosen to respond to Smith's concerns at the conference was the industry lobbyist Don Carlos Hines, who had guided the original DES New Drug Application through the FDA in 1940. Sixteen years later, Hines was still an avid promoter of the drug. "Mice are not men," he insisted; we cannot use animal studies to judge effects in humans. Instead, we should rely on the fact that thousands of women were treated with DES during menopause without apparent mishap. Hines concluded with a second familiar argument: we are exposed to natural estrogens from food and our own bodies, so even if DES residues do exist in beef, "one is not making any appreciable, or, as a matter of fact, hardly measurable, increase in the supply of estrogen that is already present."[45]

Franz Gassner, an agricultural-experiment-station researcher, reiterated the industry argument that the presence of natural estrogens in food meant that DES posed no threat to consumers. He then reassured the audience that the government was monitoring the situation closely: "It has been adequately demonstrated that with the highly sensitive control methods employed by the Food and Drug Administration, the Public Health Service, in cooperation with a host of qualified institutional investigators, are fully capable of maintaining adequate safety measures for the use of medicated feeds. Therefore, the concern expressed by the authors of the paper under discussion becomes quite pointless and unwarranted."[46] Yet as Gassner knew, the FDA was doing essentially no monitoring, which made the expert technical control he described impossible.

As concern about the harmful effects of new synthetic chemicals increased during the 1950s, the House Select Committee to Investigate the Use of Chemicals in Foods and Cosmetics, chaired by James J. Delaney, conducted a two-year inquiry into chemical additives in food. Chemicals that might leach from plastic wraps and containers into food were included in the hearings, as was the use of DES in livestock. Many researchers urged the government to require additional research on chronic toxicity and mandate more controls over additives, arguing that there was not enough evidence to show whether synthetic chemicals such as DES were safe for general prolonged use. The FDA (which had split off from the Department of Agriculture in 1940) supported some changes in the law. In contrast, the USDA remained more sympathetic to industry, with staff economic entomologists arguing at the hearings that "research . . . could go on forever without disclosing all possible hazards from expected uses."[47]

The disputes were not simply about science and safety; they were also about who held economic and regulatory power. The FDA and the USDA were competing for the right to regulate particular industries, while industry representatives and farmers' groups were challenging the right of government to regulate their activities at all. "I know of no other industry which has to comply with more laws and regulations in order to sell its product," complained Lea S. Hitchner, president of the National Agricultural Chemicals Association. Samuel Fraser, secretary of the International Apple Association, characterized the push for new laws as a "grab for power which is to be secured under the whip of hysteria." Consumer groups pointed out that without strong regulation and oversight, ordinary citizens would have no protection from the growing power of industry.[48]

Fearing that Congress might force the industry to test chemicals in food, the Manufacturing Chemists' Association hired the firm of Hill and Knowlton, strategists for the tobacco industry, to plan a response to the Delaney committee. Hill and Knowlton vigorously defended the presence of chemicals in the food supply. The company's lobbying efforts succeeded: instead of passing the proposed legislation mandating testing, Congress enacted a much weaker bill, the 1958 Food Additives Amendment, which required that the industry demonstrate to the FDA that food additives were safe. The law covered packaging material and food-related materials such as can liners and plastic wrap. Nevertheless, as the historian Sarah Vogel argues, this amendment actually weakened the regulation of risk. The earlier food laws of 1906 and 1938 had identified dangerous chemicals as hazards in and of themselves, regardless of exposure levels, and these were therefore theoretically restricted from entering food at any level. But the 1958 amendment reversed this standard "based on the logic that many of the new industrial chemicals in use, in particular the pesticide DDT, increasingly detected in cow's milk, were 'necessary in production or unavoidable.'" The amendment formally conceptualized risk as dependent on the amount of chemical exposure; consumer safety could be achieved "not by questioning the hazard *per se,* but by minimizing the exposure."[49]

The 1958 Food Additives Amendment required that any company that wanted to add a substance to animal feed had to show that it was safe for the animal. Moreover, any parts of the animal meant to be eaten by humans were not to contain chemical residues, unless those residues were safe for human consumption. If livestock producers wanted to add a toxic substance such as cement dust to cattle feed to increase weight gain, they could do so, once they demonstrated that the meat of the cow did not contain unsafe levels of residues. If producers wanted to add growth hormones to dairy cattle feed, they could do so once they had shown that any residues would be safe for human consumption. Decisions about the threshold, if any, at which residues would prove safe remained vigorously contested.[50]

Carcinogens were the exception. Delaney persuaded Congress to pass what became known as the Delaney Clause to the 1958 amendment. This had the potential to be revolutionary, for it stipulated that any substance known to cause cancer in test animals could not be added to food in any quantity whatsoever. No one needed to prove that the quantities in question caused cancer; no one needed to prove that animal studies applied to humans. Essentially, this clause repudiated the toxicology dose-response paradigm, which asserted that the dose makes the poison. Instead of assuming that if something could not be measured, it effectively did not exist, the Delaney Clause postulated that risks might exist beyond science's current power to measure or model.

The Delaney Clause was a formal, legal expression of precaution. Yet in its focus on carcinogens, it also distracted attention from other potential hazards, such as endocrine disruption. The makeup of the 1958 Food Additives Amendment reflected a schism: for carcinogens the law required almost complete precaution; for everything else it allowed a more relaxed threshold model.[51] These tidy divisions would soon be challenged, as new information about risks from chemical exposure entered the public arena in the 1960s.

CHAPTER 6

Growing Concerns

The 1960s were a turbulent era. In 1962 alone, Rachel Carson's *Silent Spring* was published, catalyzing public concern about synthetic chemicals; Congress passed the "DES proviso," which eviscerated the Delaney Clause; and the thalidomide crisis erupted, further eroding faith in public health regulation. That single chaotic year was the beginning of a decade of controversy about the environment. By 1971, the year that prenatal DES exposure was linked to vaginal cancer, federal agencies were facing intensifying demands for environmental and consumer-health protection. The use of DES in animals intended for human consumption, pesticides in the food supply, and drugs taken during pregnancy all stirred growing anxieties about synthetic chemicals with hormonal effects.

The DES Proviso

The passage of the 1958 Delaney Clause had quickly led to an industry backlash against the regulation of chemicals in food. In 1961 the National Institute of Animal Agriculture's annual conference highlighted the growing tensions between agricultural producers and the Food and Drug Administration. The vice president of Lilly thundered that the Delaney Clause was "a triumph of superstition over science. . . . If the spirit of this clause were exercised to the extreme, if we eliminated from our lives everything that causes cancer in animals, mankind would be reduced to sitting around in the dark, naked and hungry, waiting to die of cancer."[1]

Speaker after speaker sounded the same theme: although the public might be frightened, their reaction was mere emotion to be dispelled by scientific expertise. To fight Cold War threats to the free market, scientists, industry, and government had to join together to reassure consumers. As the president of the American Meat Institute, Homer Davison, proclaimed, "Fundamentally we all believe in one basic philosophy. That philosophy accepts the free market system as a base. From there it argues that by increasing the production of animal protein foods, the health of agriculture and the health of people can and will be improved. Thus our thesis is very simple."[2] More meat meant a healthier nation, which would support a healthier free-market system. Hormones could make this happen.

By 1962 livestock producers had gained enough support in Congress to modify the Delaney Clause with the DES proviso, an exemption that permitted the use of DES in livestock meat production provided no detectable residues were found in the edible tissue. The drug companies assured the FDA that they had done numerous tests that proved that all traces of the drug would be gone after forty-eight hours. On these grounds, the FDA agreed to allow the use of DES for livestock, even though the agency knew that it lacked the staff or funds to monitor compliance with the proviso.[3]

Permitting a chemical to be used in livestock feed as long as no residues end up in the meat may seem like a reasonable compromise. But to regulate the chemical's use based on this constraint assumes several things: first, that tests exist that can detect biologically relevant levels of residues and that those tests are reliable and reproducible; second, that someone is doing the testing; third, that the farmers are following exactly the set protocols for giving the animals the right amount of the chemical and then withdrawing it from the feed; fourth, that a small sample of meat on the market is a reliable indicator of all the meat on the market; and finally, that the animals submitted for sampling are being treated in the same way as other animals. All these assumptions rest on the fundamental belief that even without strong government oversight, industry will act in the best interests of consumers, rather than stockholders. While the FDA Commissioner, George Larrick, believed this wholeheartedly, little evidence from the agency's two decades of experience with DES supported it.

Rachel Carson and *Silent Spring*

The publication of Rachel Carson's *Silent Spring* in 1962 ignited a controversy over chemical residues in food and the environment that raged throughout the decade. Carson fostered a growing public conversation about the dangers of synthetic chemicals, particularly DDT (which later turned out to have estrogenic effects similar to DES). Carson did not just expose the risks posed by pesticides; she also challenged a worldview that insisted science could and should master nature.

Carson recognized that World War II had helped create an ideology of technological control over the unruliness of nature, a concept that found expression in the adoption of synthetic pesticides in field crops and synthetic hormones in pregnant women and cattle. As the historian Edmund Russell argues in *War and Nature,* agricultural experts called for "total war against man's insect enemies, with the avowed object of total extermination instead of mere 'control.'" In this new conceptual framework, complete control of insects was desirable, and victory over agricultural foes became as important as victory over military enemies. Wartime publicity campaigns had promoted the idea that the new chemical wonders could solve all insect problems—on the farm, in the suburban landscape, and in the home.[4]

The war had created new institutional structures for scientific funding and research, leading to powerful alliances between civilian university scientists, armed forces, and industry. The stunning practical power of science demonstrated during World War II had pushed researchers to develop ways to apply the products of war to agriculture. Specifically, the enormous postwar growth in the use of DDT and similar pesticides came about because of economic ties between the military and the chemical industry.[5] Chemical companies had used the capital and expertise gained during the war to expand insecticide research, and scientists had promoted the concept that human beings were engaged in a battle for survival with insects, helping to shape the ideology of control over nature that Carson deplored.[6]

A central focus of *Silent Spring* was DDT, for Carson recognized that it eluded traditional methods of detecting risk. Like many other chemicals developed during the war years, DDT was originally envisioned as a miracle chemical that would improve the quality of human lives with relatively little risk to people. At the beginning of World War II, diseases were killing more soldiers than weapons. In particular, lice carried typhus, and before 1942 louse-control technology had relied on pyrethrums from plants that needed to be imported from Africa. The war had disrupted the international trade networks that supplied pyrethrums to the Allies, and a substitute was desperately needed. DDT proved to be the answer. In November 1942 the USDA learned that DDT killed lice and appeared to be nontoxic to people. In late 1943, when typhus broke out in Naples, Allied health organizations dusted more than a million civilians with DDT powder. For the first time, public health actions managed to halt a winter typhus epidemic. The success, in Russell's words, "lifted hopes for 'total victory' against other insects—not just a relationship of uneasy balance, but total annihilation."[7] Researchers in the United States found that extremely low doses of DDT killed mosquitoes, not just lice, and the Allies began spraying DDT where soldiers were threatened by malaria.

The spraying of DDT led to great health benefits for soldiers. The Swiss chemist Paul Müller, who had discovered the insecticidal properties of DDT, was awarded the 1948 Nobel Prize for his work on DDT. Yet not all military scientists saw the chemical as a wonder drug. Early tests raised concerns about DDT's safety for people, lab animals, and birds. Further tests, however, showed that skin absorption did not seem to result in the same kind of toxic reaction, and although researchers did not conclude that DDT was harmless, they did decide that the hazards needed to be weighed against the advantages for military use. Soldiers would be exposed for short durations, so the risks seemed reasonable. After the war, however, civilians were exposed for very long durations, a situation that required a different calculus of risk and benefit.

Military and federal agencies alike expressed concern about DDT's potential for harm when broad populations were exposed. In 1944 scientists at the FDA warned in a chemical industry journal that DDT was toxic when fed to animals, even at very low doses, indicating that "the safe chronic levels would be very low indeed." Other FDA staff soon announced that DDT fed to dogs accumulated in body fat and breast milk. Federal scientists were particularly concerned over DDT in milk, because goats fed DDT produced milk toxic enough to kill the rats that drank it. When DDT showed up in cow milk destined for human consumption, federal entomologists recommended that dairy managers switch to other pesticides, concerned that children might suffer the same fate as lab rats.[8] The properties that made DDT so useful in the war—persistence and a broad spectrum of activity—were the same traits that caused concern within the scientific community about its potential health effects for consumers.

In addition, ecologists worried about broader ecosystem-level effects of the chemical because wartime DDT spraying in the South Pacific islands had seemed to devastate some island ecologies. Early in the chemical's military use, entomologists feared that DDT could worsen pest problems by killing off competitors that kept those pests in check. Field trials in Panama suggested DDT could create biological deserts devoid of life. In May 1945, Clarence Cottam of the Fish and Wildlife Service urged that DDT not be released for civilian use until the service could better assess its ecological effects. Other scientists recommended that aerial spraying of DDT be reserved for military emergencies. Yet as Russell describes, these warnings went unheeded, and the chemical made its way out of military control and into public use.[9]

By the 1960s public reaction against DDT had begun, stimulated by fears of aerial spraying. In the early 1960s, more than a hundred million acres of crops—one-sixth of the cultivated lands in the nation—were sprayed with pesticides. Crop dusters were able to treat vast landscapes quickly and efficiently from the air, but drifting chemicals crossed property boundaries and began to affect suburbanites, creating a substantial backlash against the chemicals. Homeowners often used toxic chemicals in their own houses, but at least they felt that they controlled the decisions about whether and how much to spray. Aerial spraying, by contrast, was out of the homeowners' control and so threatened their sense of home as a refuge; it introduced what the philosopher Carl Cranor calls "toxic trespass." When consumers realized that they had little control over the substances that entered their homes and bodies, they joined with scientists in challenging DDT spraying. As the historian Pete Daniel argues in *Toxic Drift,* the Department of Agriculture blocked efforts to protect public safety even when medical evidence of pesticide harm accumulated. Instead, the agency marshaled its resources to support agricultural interests.[10] The Food and Drug Administration, however, found itself caught in the middle. Rather than deny that consumer harm could result from pesticides, the FDA insisted that scientific control would minimize the risks, keeping them under reasonable thresholds. Carson's arguments in

The great expectations held for DDT have been realized. During 1946, exhaustive scientific tests have shown that, when properly used, DDT kills a host of destructive insect pests, and is a benefactor of all humanity.

Pennsalt produces DDT and its products in all standard forms and is now one of the country's largest producers of this amazing insecticide. Today, everyone can'enjoy added comfort, health and safety through the insect-killing powers of Pennsalt DDT products . . . and DDT is only one of Pennsalt's many chemical products which benefit industry, farm and home.

GOOD FOR STEERS—Beef grows meatier nowadays . . . for it's a scientific fact that—compared to untreated cattle—beef-steers gain up to 50 pounds extra when protected from horn flies and many other pests with DDT insecticides.

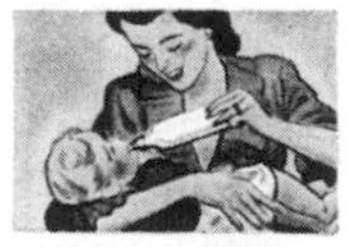

Knox Out FOR THE HOME—helps to make healthier, more comfortable homes . . . protects your family from dangerous insect pests. Use Knox-Out DDT Powders and Sprays as directed . . . then watch the bugs "bite the dust"!

Knox Out FOR DAIRIES—Up to 20% more milk . . . more butter . . . more cheese . . . tests prove greater milk production when dairy cows are protected from the annoyance of many insects with DDT insecticides like Knox-Out Stock and Barn Spray.

GOOD FOR FRUITS—Bigger apples, juicier fruits that are free from unsightly worms . . . all benefits resulting from DDT dusts and sprays.

GOOD FOR ROW CROPS—*25 more barrels of potatoes per acre* . . . actual DDT tests have shown crop increases like this! DDT dusts and sprays help truck farmers pass these gains along to you.

Knox Out FOR INDUSTRY—Food processing plants, laundries, dry cleaning plants, hotels . . . dozens of industries gain effective bug control, more pleasant work conditions with Pennsalt DDT products.

CHEMICALS

97 Years' Service to Industry • Farm • Home

PENNSYLVANIA SALT MANUFACTURING COMPANY

WIDENER BUILDING, PHILADELPHIA 7, PA.

*An advertisement for commercial DDT (*Time, *June 30, 1947)*

Silent Spring, however, made many consumers believe that scientific control was not possible.

Carson gave voice to growing concerns about the technocrats' concepts of risk. DDT emblematized both the new risks posed by synthetic chemicals with hormonal actions and the difficulties regulators had in responding to these risks. When consumers and scientists had expressed concerns about widespread application of the chemical, regulators and industry lobbyists had countered with what the historian Thomas Dunlap described as "DDT's apparently perfect safety record." They insisted that the chemical was harmless: "No one died from DDT poisoning; no one even got sick. Long technical discussions of chronic toxicity, lesions on rat livers, and parts per million in milk did nothing to disturb this presumption."[11] The complexities of DDT stemmed from the fact that it was not acutely poisonous to people or most wildlife; rather it was a chemical with hormonal mechanisms of action that could not be easily understood or contained within traditional paradigms of risk. Few scientists had initially suspected that the chemical's most damaging effect might be disruption of the endocrine system rather than immediate death. But even as many scientists and regulators were insisting that DDT was relatively harmless, other scientists were becoming aware that synthetic chemicals that disrupted hormonal functions might pose unique risks. In the early 1960s wildlife biologists had noticed a decline in falcon populations in Europe and North America. After a great deal of interdisciplinary research, biologists eventually connected bird population declines to eggshell thinning, a hormonally modulated effect of DDT. Because DDT changed estrogen levels and eventually led to reproductive failure, exposure to the chemical could conceivably lead to the collapse of an entire peregrine falcon population without killing a single adult.[12]

Other hormonal activities of DDT were also becoming apparent. In 1950, the biologists Howard Burlington and Verlus Frank Lindeman had found that DDT effectively castrated male chicks: the testes of treated chicks were less than one-fifth the size of those of control chicks. Other researchers discovered that DDT could alter the formation of enzymes in the liver; this in turn would alter the formation and regulation of estrogen, progesterone, and testosterone, affecting reproduction. Journalists were fascinated by DDT's potential to affect sexual traits. Headlines such as "Scientist Warns of DDT Peril to Sex Life," "Scientist Fears DDT Can

Cause Sex Change," and "DDT Termed Peril to the Sex Organs" were rife, foreshadowing the media attention that greeted endocrine-disruption research a quarter century later.[13]

Rachel Carson had warned of the ecological effects of pesticides and the links between humans and wildlife, making a crucial observation when she noted that DDT might have transgenerational effects: "The insecticidal poison affects a generation once removed from initial contact with it."[14] Researchers such as John McLachlan at the National Institutes of Health began using DES as a surrogate to help understand DDT's transgenerational effects, which led to the discovery that the two chemicals had close similarities in their ability to disrupt fetal development. Growing concerns about transgenerational effects focused on the key issue at stake with DDT, DES, and other hormonally active chemicals: prenatal exposure was difficult to monitor, but it might pose the greatest risk.[15]

Thalidomide

The transgenerational effects of synthetic chemicals that Carson warned about in 1962 took center stage that same year with the thalidomide crisis. Drug regulators were horrified to discover that thalidomide, a sedative that was Europe's best-selling drug after aspirin, was causing terrible birth defects in children whose mothers had taken it while pregnant. Although the number of affected children had been limited in the United States thanks to FDA staffer Frances Oldham Kelsey, who had resisted industry pressure to approve the drug, consumer advocates realized that no U.S. regulations existed that would prevent another such tragedy. These concerns led Congress to pass the Kefauver-Harris Drug Amendment in 1962, for the first time requiring that before a new drug could be used on women of childbearing age, litter tests needed to be done on animals to detect adverse effects during pregnancy. Yet despite the anxiety sparked by the thalidomide crisis and *Silent Spring,* FDA regulators continued to resist pressure to restrict DES. The thalidomide case can help us understand how the FDA dealt with concerns over the prenatal effects of chemicals, as well as why DES escaped closer scrutiny for nearly a decade afterward.

A European company, Chemie Grünenthal, first synthesized thalidomide in 1954 when it was trying to develop low-cost antibiotics. The drug turned out to have no antibiotic properties, but Chemie Grünenthal patented it anyway and began to look for a disease that thalidomide could cure. Early laboratory tests suggested that even high doses of thalidomide would not kill rats, mice, rabbits, cats, or dogs, which made the drug's commercial prospects attractive if the company's scientists could discover a use for it. In the journal *Pediatrics* the physician William Silverman argued in 2002 that even though the drug did not act as a sedative in experimental animals, the company knew that a nontoxic sedative would have enormous commercial potential, so it "decided to explore the possibility that the new agent might act as a sedative in humans. Early in 1955, with no scientific rationale and with no formal plan to monitor immediate results or to conduct a systematic follow-up, Grünenthal distributed free samples of the new drug to doctors in West Germany and Switzerland."[16]

When patients reported "soothing, calming, and sleep-inducing effects," the manufacturer knew it had found a market for the drug. Silverman noted that on Christmas Day 1956, ten months before the company placed the drug on the German market, "a child was born without ears. The father, an employee of Chemie Grünenthal, had brought home samples of the new drug for his pregnant wife; years later, he learned that his daughter was the first living victim of the subsequent epidemic of thalidomide-induced infant deaths and malformations." Yet the company claimed the drug was completely safe, and thalidomide was sold over the counter as a sedative. Because the drug seemed to prevent nausea in pregnancy, Chemie Grünenthal was determined to make pregnant women a significant share of the market. "In August 1958, Grünenthal wrote a letter to all German physicians declaring that 'thalidomide was the best drug for pregnant and nursing mothers,'" even though the company had not conducted a single study of drug's fetal effects. Sales of thalidomide soared.[17]

The Cincinnati-based Richardson-Merrell Company bought the U.S. license for the drug, and on September 12, 1960, the company submitted a New Drug Application to the FDA requesting approval for thalidomide. Because the drug was already widely used in Europe, the agency expected approval to be routine, and so assigned the case to Kelsey, who was new to the FDA. As Kelsey later recalled, "Since I was new, they selected an easy one for me." But the case turned out to be anything but easy. Kelsey had a great deal of experience with drug research, and that experience had made her skeptical of industry claims. She had earned her Ph.D. in 1938 at the University of Chicago in pharmacology, studying with E. M. K. Geiling, one of the most renowned pharmacologists of the era. She then joined the faculty at the University of Chicago and earned her M.D. while carrying out research on synthetic anti-malarial drugs. Her research showed that anti-malarial drugs could cross the placenta, something few researchers had suspected was possible. While in Chicago, she also served on the editorial staff at the *Journal of the American Medical Association,* an experience that made her leery of industry research. As she recalled, "When I came to the Food and Drug Administration . . . , I found that many of the studies in support of safety of the new drugs were done by investigators whose work had not been accepted for publication in the Journal."[18]

After reviewing the materials Richardson-Merrell had submitted, Kelsey refused to approve the application. Two issues particularly concerned her: what she called the drug's "curious lack of toxicity," which made her suspicious about the quality of the company's data, and the lack of good evidence on its relation to metabolism and excretion. She doubted the European parent companies knew much about how the drug worked in the body, certainly not enough to claim that it was safe for pregnant women.[19]

When Kelsey rejected the New Drug Application for thalidomide, the FDA allowed Richardson-Merrell to gather more data and resubmit the application. The company's medical director, Joseph Murray, campaigned hard to push the application through the FDA, repeatedly calling and visiting the agency headquarters. The pressure on Kelsey intensified during the spring and summer of 1961: "They came to Washington, it seemed, in droves. They wrote letters and they telephoned—as often as three times a week. They telephoned my superiors and they came to see them too. . . . Most of the things they called me, you wouldn't print." Murray even contacted Larrick directly, asking him to remove Kelsey from the case.[20]

In the midst of intensifying industry pressure, Kelsey came across a report in the *British Medical Journal* suggesting that thalidomide might cause peripheral neuritis, a type of nerve damage in the hands and feet. This article had appeared before Richardson-Merrell resubmitted its application, yet the company had failed to report it to the FDA. Kelsey immediately questioned Richardson-Merrell staff members about this, and they claimed that they had only recently seen the article. Even though the staff insisted that the British findings were insignificant, Kelsey believed that they raised a red flag about the drug's possible effect on developing fetuses, which might be exposed if pregnant women took the drug for nausea. In her earlier research, Kelsey had witnessed the ways anti-malarial drugs were metabolized by the liver, noting that fetuses broke down drugs differently from the way adults did because "the baby simply doesn't have the enzymes to protect itself . . . as the adult did."[21]

In November 1961, reports began to emerge from Europe that mothers who had taken thalidomide during pregnancy were giving birth to babies with severe defects. Kelsey reported that "children had been born with hands and feet protruding directly from their torsos. Others had limbless trunks with toes extending from their hips; others were born with just a head and a torso; still others had cardiac problems." A pediatric cardiologist, Dr. Helen Taussig, traveled to Europe to investigate the reports. Upon her return, she testified before the Senate, helping Kelsey prevent the approval of thalidomide in America.[22]

Larrick was slow to respond to Kelsey and Taussig's concerns about thalidomide. As the physician and historian Jack C. Fisher writes, "Not until July 23, 1962, eight months after Merrell informed the FDA that thalidomide had been withdrawn from the German market, did Commissioner Larrick dispatch inspectors to Cincinnati where they uncovered a series of deceptions that would soon shock the government, the American public, and especially the legitimate pharmaceutical industry. When Chemie Grünenthal had begun its search for an American licensee, records from their own experiments already documented unacceptably high mortality rates among animals fed thalidomide. When Murray had visited Germany and England soon after the reports of neuritis appeared, the company's files held more than 400 letters describing patients with a variety of central and peripheral neurological symptoms."[23] Kelsey believed that deception was involved in the drug company's failure to report adverse effects. Richardson-Merrell, however, argued that the company had not reported the evidence that the drug was dangerous to the FDA because no law required that foreign data be reported. Instead, the FDA received only data from the company that claimed to show the drug was safe during pregnancy. Kelsey suspected that these tests had been conducted inadequately or that the results had simply been faked.[24]

In the midst of the scandal, Larrick was "permitted to retire quietly," and Kelsey was granted the President's Award for Distinguished Federal

Dr. Frances Oldham Kelsey receives the President's Award for Distinguished Federal Civilian Service from President John F. Kennedy, 1962 (photograph courtesy of the National Library of Medicine, "Images from the History of Medicine Collection," A018057)

Civilian Service, the highest honor possible for a U.S. civilian. Yet even though Kelsey's efforts kept the drug off the market, tens of thousands of Americans had still been exposed to thalidomide. Under the guise of performing the safety tests necessary for the FDA approval, Richardson-Merrell had marketed the drug widely in the United States to pregnant women. While Kelsey's team was reviewing the New Drug Application, Richardson-Merrell distributed thalidomide to more than twelve hundred U.S. physicians, claiming it was for "investigational use," even though the company's sales and marketing division organized the distribution. More than 2.5 million pills were given to more than twenty thousand patients during these trials. Richardson-Merrell never informed most of the exposed women about the risk of birth defects.[25]

For all the public attention paid to thalidomide, the crisis did not lead to closer scrutiny of the prenatal effects of DES. By the early 1960s, research, medical, and regulatory communities had received warnings that chemicals could harm fetal development. Thalidomide was not the only chemical of concern; Kelsey warned doctors that birth defects and deformities were increasing, and she was troubled by DES and other chemicals with hormonal activity. Thalidomide produced immediate, massive birth defects, yet with other chemical exposures problems did not emerge until puberty, making it almost impossible to link fetal exposure to later harm. Kelsey warned that "perhaps 50% of congenital malformations are not recognizable at birth." Yet even though reports were appearing in the technical literature about apparent intersex conditions in children exposed to DES in utero, few doctors other than Kelsey could imagine that a chemical given during pregnancy might have effects that would only emerge decades later.[26]

Even Edward Charles Dodds, the British biochemist who first synthesized DES, did not make the connection between thalidomide and DES. In a 1971 lecture Dodds told his audience, "I must say that when the Thalidomide tragedy occurred I could not help feeling how lucky we were that there was no hidden toxicity in stilboestrol." Dodds noted how concerned he had been about the potential health effects of long-term treatment with the synthetic hormones he had helped discover: "I personally as a biologist cannot believe that it is possible to interfere in a natural process for so long without something really serious happening." Ironically, at almost the same time Dodds was speaking, researchers in Boston were finding that vaginal adenocarcinoma in young women was associated with prenatal exposure to DES.[27]

Why didn't Frances Kelsey express more concern about DES's effects on pregnancy, given her experience with thalidomide? Kelsey knew that DES had never been sufficiently tested on pregnant animals in the laboratory, and she was also aware that steroid hormones, when given to pregnant women, could cause masculinization and sexual alteration of the fetus. In one talk, she noted that "the teratogenic effect of certain progestational substances and other steroids advocated for the control of bleeding in pregnancy, or the treatment of threatened abortion, was first described in 1958. Such drugs given during the first trimester of pregnancy lead in a small percent of cases to masculinization of the external genitalia in the female fetus." In another talk, Kelsey mentioned a researcher named Wilkins who in 1958 had described occurrences of pseudohermaphrodism in female infants whose mothers had taken synthetic progestins while pregnant. In 1963, Kelsey warned that diethylstilbestrol might have similar effects. She told physicians that "certain progestins used in early pregnancy for threatened abortion may give rise to masculinization of the external genitalia in the female fetus. . . . Similar effects may follow testosterone preparations or large doses of diethylstilbestrol."[28]

Kelsey's concerns about growing chemical exposures never led to her to publically call for a ban on DES, but she did insist that regulators not assume that drugs and chemicals were safe until proven otherwise. The government had a responsibility to require evidence of safety before approval, to monitor effects after approval, and finally to respond to the results of that monitoring.[29] Her call for increased regulatory oversight of drug companies helped shaped the 1962 Kefauver-Harris Drug Amendment, which for the first time required that a drug marketed for a particular purpose had to work better than a placebo. The Kefauver-Harris Drug Amendment instituted a drug-approval process that included orderly clinical trials, which prevented manufacturers from widely distributing the drug before approval. In addition, drug companies had to report to the FDA any adverse reactions to the drug that appeared in tests. Most important, potential human subjects had to give their informed consent before being given the drug.

The amendment also required that all existing drugs be reviewed for efficacy as quickly as possible to ensure they conformed to the new standards. The FDA contracted with the National Academy of Sciences to perform the evaluations on drugs that were already on the market. After five years, in 1967 the academy panel reported that DES was essentially ineffective, writing that "its effectiveness cannot be documented by literature or its own experience." Yet the panel did not explicitly recommend that DES be withdrawn, noting that many doctors believed the drug was useful, even if medical evidence did not prove its efficacy or its safety for fetuses. After the panel's report, the FDA could legally have withdrawn approval for the drug, yet the agency chose not to do so, a decision that troubled consumer advocates, who were aware that the drug caused cancer in laboratory animals.[30]

In 1967, while the FDA was considering whether to restrict the use of DES for women, the agency was continuing to struggle with the problem of DES residues in beef. The FDA was responsible for the drug's regulation, yet the USDA was responsible for testing meat for residues. The testing protocols used by the USDA were capable of detecting DES only at levels of 10 parts per billion, even though FDA staff knew that DES caused tumors in mice at 6.5 parts per billion in feed. A slab of meat found by the USDA tests to have "no detectable residue" might still have residues capable of causing rodent tumors. The USDA sampled only six hundred cattle each year out of the tens of millions treated with DES, and even with the insensitive assays, many samples were contaminated with illegal levels of residues. In 1965, 0.7 percent of cattle sampled had illegal residues of DES, and those numbers rose to 1.1 percent in 1966 and 2.6 percent in 1967.[31]

These were clear violations of the law and evidence that the DES proviso was not working, for millions of cattle were indeed showing up in the nation's food supply with residues of a carcinogen. Yet rather than banning the chemical, the FDA and the USDA agreed that the problem lay with cattle growers who were not obeying the regulations, rather than with faulty assumptions about technological control embedded in the DES proviso.[32]

In 1970, consumer groups, infuriated at what seemed to be the agency's refusal to apply the law to DES, pressed for congressional hearings, even as new information about DES's effects on women was coming to light. Representative Lawrence Fountain of North Carolina scheduled congressional subcommittee hearings on the problem of DES in agricultural feeds for March 1971. Shortly before the hearings were convened, the FDA learned that DES had been linked to cancer in humans, a discovery that shattered the assumptions underlying the agency's insistence that the chemical was safe. But instead of revealing these findings to Congress, the FDA withheld them, a decision that when discovered only lessened public confidence in the agency.

Diethylstilbestrol and Cancer in Women

In April 1971 a groundbreaking article appeared in the *New England Journal of Medicine.* For the first time, prenatal exposure to a chemical was shown to have caused cancer in humans. That chemical was diethylstilbestrol, and the finding reverberated throughout the medical world and the agricultural industry.

Two years earlier, Dr. Arthur Herbst and his colleagues at Massachusetts General Hospital in Boston had been surprised to find that seven young women with an extremely rare cancer (clear-cell adenocarcinoma of the vagina) had been treated at the hospital between 1966 and 1969. In the medical literature worldwide, only four cases of this cancer had ever been reported in women under the age of thirty. The doctors asked the epidemiologist David Poskanzer to help them search for possible links between the women, just as one more young woman appeared in Boston with clear-cell adenocarcinoma.

As the medical historian Diane Dutton recounts, the researchers set up a case-control study, searching out women of similar ages and backgrounds who had not developed the cancer. From 1969 to 1971 they searched for clues, interviewing both sets of women (the patients and the controls) extensively, searching for differences in their habits that might illuminate why some of the women developed cancer but others did not. Researchers asked the young women about smoking, douching, birth-control pills, family histories, even tampons. Much to their frustration, they could not find any significant differences between the control women and the women with cancer. A breakthrough finally came in early 1971 when one of the patients' mothers urged the doctors to consider the stilbestrol pills she had taken during pregnancy. The researchers compared the medical records of the cancer patients' mothers, learning that seven of the eight had taken diethylstilbestrol during the first trimester of pregnancy.[33]

Herbst and his colleagues immediately submitted a report to the *New England Journal of Medicine,* the leading American medical journal. Rather than wait until scheduled publication in April, the editor immediately alerted the FDA, sending the agency prepublication galleys of Herbst's article in March 1971. Fountain had already scheduled subcommittee hearings on DES in agricultural feed for that March, well before he learned about Herbst's still-unpublished findings. At the March hearings, Fountain had still not heard about Herbst's findings, but the FDA staff who testified at the hearings had. Nevertheless, at that hearing, FDA commissioner Charles C. Edwards and Henry Simmons, director of the Bureau of Drugs within the FDA, assured the committee that "they knew of no instance of human cancer caused by DES in the 30 years the drug had been administered to men and women." Commissioner Edwards, a physician with close ties to the pharmaceutical and tobacco industries who had been appointed by President Nixon to head the agency, was almost certainly lying, for he had already been personally informed of Herbst's findings. Further, the FDA did not inform doctors, farmers, consumers, or drug companies about the new findings, allowing the drug to remain in use.[34]

Other public health agencies did act, however. Dr. Peter Greenwald, the director of the Cancer Control Bureau of the New York State Department of Health, read Herbst's article in April and was stunned. He searched New York's cancer registry for reports of the rare cancer and quickly found five more cases of clear-cell adenocarcinoma of the vagina, all in young women whose mothers had been given a synthetic estrogen during pregnancy. The *New England Journal of Medicine* published his results immediately, for they corroborated Herbst's findings. The New York State commissioner of health, Hollis Ingraham, responded as well, writing to all thirty-seven thousand practicing physicians in New York to warn them of the dangers of estrogen administration during pregnancy. Ingraham also wrote to Edwards in June, even before Greenwald's work was published, urging the federal agency in the strongest possible terms to ban DES for use during pregnancy. A staff member acknowledged the letter, but months passed before anyone within the agency asked the New York officials for more information.[35]

Officials in New York soon found sixty-two more cases of adenocarcinoma of the vagina or cervix in young women. Greenwald forwarded these reports to the FDA, but as had happened with Herbst, months passed before he heard anything more from the FDA. Medical staff in both New York and Massachusetts tried to determine which women had used DES so that they could be informed and their children's health could be monitored for possible cancer and other reproductive issues. The staff soon realized that without a national registry, such efforts would be wasted. Greenwald and Herbst urged the FDA to create a national registry to aid this effort, but they received no response.[36]

The *New England Journal of Medicine* published a special editorial pointing out the powerful association between cancer and DES use, and then went on to make explicit the concern this raised about DES residues in beef: "Of 40,000,000 cattle slaughtered in this country each year, 30,000,000 have been fed stilbestrol to increase their weight. . . . Since the fetus is so much more vulnerable to minute doses of a carcinogen, there is no way of judging the risk of stilbestrol residue that remains undetected by the current government assay method."[37]

After this editorial, members of Congress grew increasingly frustrated with the FDA's lack of response. Fountain called another set of hearings on DES in beef in November 1971, and this time he was unwilling to believe the reassurances of Commissioner Edwards. Immediately after the March committee hearings, when Fountain had learned of the vaginal cancer link, the Department of Agriculture had assured him and other members of Congress that meat from DES-treated cattle was certain to be safe for no residues had been found in sampled meat. The National Resources Defense Council had discovered that this assurance was a lie, for the USDA had indeed found substantial residues in tested meat at levels of up to 37 parts per billion, high enough to cause cancer in laboratory animals. When confronted with the evidence, the USDA assistant secretary had apologized to Congress, but the damage had been done. Confidence collapsed in the ability of the USDA and FDA to regulate DES.

The November Fountain hearings reflected this change in tone. Fountain and his staff members attacked Edwards with vigor. When Edwards was asked why he had delayed banning DES or warning anyone about it after the cancer connection had come to light, he replied that he did not wish to harm the industry. Fountain berated Edwards for his delays, and in response Edwards criticized the New York officials for warning physicians to stop prescribing DES in pregnancy, claiming that they had been "premature" in their actions. Fountain was incredulous at this response, pointing out that the FDA's obligation was to protect public health, not industry profits.[38]

The FDA drug chief, Simmons, spoke next, insisting that no one had fully proven, beyond any doubt, that DES caused vaginal cancer, and so sound science required further study before action could be taken.[39] When asked what studies would be needed to prove causation, Simmons outlined a program of study that would take decades, if not centuries: "There would be larger studies similar to what Herbst has already started. There would be confirmatory studies of what he has already reported. There would be studies in which we would go back and identify and then follow mothers." And finally, Simmons continued, the FDA would need to study the "benefit-risk ratio" for DES before taking action. At this point, a frustrated congressional staff member accused Simmons of having created impossible standards for demonstrating risk and placing the burden of proof on the consumer, rather than the industry.[40] The Food and Drug Administration had, in effect, reversed the precautionary principle the agency had once articulated.

Arguments over what constituted clear evidence of harm accentuated the growing distance between the FDA bureaucrats who headed the agency and scientists both outside and within the agency. Peter Greenwald testified that he believed that DES was indeed responsible for cancer in the women he observed, even if the FDA had not yet received what agency bureaucrats considered absolute proof. Greenwald pointed out that absolute proof linking chemical exposure to cancer in humans could never exist because direct experiments on human subjects were unethical and illegal. He added, "I think epidemiologically, the evidence is about as strong as you can get. We cannot do experiments but we have the fact of this clustering within a short period during the time when the drug was widely used and the fact that we had no instances of people with cancer that did not have the synthetic estrogen. . . . So I think there is strong evidence of a cause and effect relationship."[41] For a cancer researcher, this was a powerful statement indeed.

Edwards and Simmons, by contrast, prepared statements that used the same data as Greenwald's testimony, yet presented in a such a way as to suggest to an uninformed audience that the association between cancer and DES was weak. When Fountain asked Edwards whether DES was a human carcinogen, Edwards replied, "There is no evidence that DES or any other estrogen produces cancer in humans when these drugs are administered in low dosages appropriate for endocrine replacement therapy." This may have been true, but it hardly addressed the question. Edwards admitted that some men who were treated with DES for prostate cancer developed cancer of the breast, but he stressed that these findings did not constitute absolute proof. As his critics noted, if they were not strong evidence of an association, it would be hard to imagine what would be.[42]

Fountain called his November hearings specifically to revisit the vexed question of DES in livestock. Just before the hearings opened, on November 1, the National Resources Defense Council had filed a complaint in a U.S. district court requesting the court to "declare illegal the use of DES in cattle and sheep raised for human food." Now that the hormone was in litigation, the residue questions became all the more contentious. Edwards and Simmons both admitted that illegal residues had been found (and lied about), but they claimed that any DES residues in beef were of trivial importance, so their failure to uphold the law was not really an issue.[43]

Commissioner Edwards made several key claims borrowed directly from industry to defuse the attacks on DES. First, DES was no different from natural estrogens. In addition, because some groups of people already had high levels of estrogen in their bodies, and because some foods contain natural phytoestrogens, the extra estrogens from DES residues in beef were surely meaningless. Finally, animal studies were not human studies, and because no one had ever proven that DES caused cancer in humans, there was no need to act.[44]

Fountain called Dr. Roy Hertz to respond to the FDA claim that DES was safe in beef, and it was his testimony that was most damaging to the FDA's case. Hertz had been with the National Institutes of Health for twenty-eight years, and he had spent much of that time consulting for the FDA. One of the world's leading cancer researchers, he trained in medicine, public health, and physiology, but he also had experience in livestock physiology and endocrinology. Hertz's testimony undermined each of the claims Commissioner Edwards had made about the safety of DES residues.

Hertz began by arguing that low doses of DES might actually be more problematic than high doses, depending on the timing of exposure. One short, intense exposure to DES, similar to the exposure a woman receives to her own body's estrogens during pregnancy, might not cause severe harm, but a much lower exposure over a longer period of time might cause harm.[45] Hertz pointed out that in animal studies the dosage of exogenous estrogen needed to induce cancer could be extraordinarily tiny, even when the body was producing its own endogenous estrogens, which were not causing cancer. The fact that bodies produce their own natural estrogens did not mean that constant exposure to synthetic estrogens was safe. As Hertz noted, "We do know that if we disturb the [estrogen] balance by various physiological experiments thereby doubling what the body produces itself, we can, under certain experimental conditions, elicit malignant changes in the peripheral tissue. . . . Now, we do not know whether the additional load added by the amount which is proposed and identified in animal meats . . . represents a comparable disproportion in the endogenous load when this material is taken daily, constantly, throughout the individual's life."[46]

There are two ways of looking at the burden of synthetic estrogens in relation to natural estrogen loads. From one perspective, if natural estrogens are known carcinogens, adding more estrogens would make little difference since human bodies already produce natural estrogens and the human race has survived them. From the other, if natural estrogens cause cancer, it would be prudent not to add any further unnecessary cancer-causing chemicals to the body. As Hertz put it, "We carry in our daily food intake of natural foods a certain estrogenic dietary load. It does not seem to me that, under those conditions, together with the endogenous load, it is wise to increase the intake of a substance which leads to a certain proportion of cancers. . . . What sense does it make to add an additional load at will? None whatsoever in my estimation."[47]

Decades of animal experimentation had provided evidence that DES caused cancer, Hertz insisted, but although "all of the known human carcinogens also produce cancer in animals," doctors often ignored animal evidence and substituted their anecdotal experiences with DES. These doctors may have convinced themselves that DES was safe, but Hertz pointed out that an intuitive sense of safety was no substitute for epidemiological evidence. Doctors had the same sense of intuitive safety about cigarettes, Hertz noted, and the tobacco industry manipulated this belief to dismiss findings from animal experiments. Hertz stressed the parallels between tobacco and estrogens: "Our inadequate knowledge concerning the relationship of estrogens to cancer in women is comparable with what was known about the association between lung cancer and cigarette smoking before extensive epidemiological study delineated this overwhelmingly significant statistical relationship." (One wonders what Commissioner Edwards, who had ties to the tobacco industry, thought about Hertz's choice of analogies, for Edwards was being no more diligent about regulating tobacco than he was about regulating DES.) Hertz testified that Herbst's findings of vaginal cancer in women exposed prenatally to DES should finally convince the FDA, industry, and physicians that animal studies do indeed have relevance for humans.[48]

The most telling of Hertz's arguments was his response to the FDA commissioner's claim that he could address the residue problem by implementing better enforcement of existing regulations. After a career in government, Hertz had lost his faith in the ability of the FDA to enforce its own regulations. As he testified: "From my own personal knowledge over the years, having studied the poultry practices and having studied the beef practices and having observed related problems of hormonal contamination . . . we are dealing with a class of substances here the order of potency of which is so great that their physical control in our environment needs the most rigorous steps. . . . What has been proposed in the reports from the Food and Drug Administration I believe is unfeasible and impractical and ill-advised." He went on to tell of the problems that cropped up "under laboratory-controlled conditions, where occasionally the presence of estrogen-treated animals in the laboratory leads to accidental contamination. . . . A case in point was the almost universal contamination of all the animal food supplies used in rat experiments up and down the east coast some years ago due to contamination with estrogens from one . . . milling plant." If control could not be assured in the laboratory, it certainly could not be expected in the feedlot, in the slaughterhouse, or on the dinner plate.[49]

Hertz closed by expressing his concern about the possibility that society might be poisoning not just people but entire ecological systems: "I have not even touched on all of the problems of the potential additional contamination of the handlers as well as the physical environment of the animals excreting large amounts of this material into the soil and into the water supplies in localized areas, a subject which has not received the attention that it deserves."[50] In 1958, Hertz warned that "the fecal excretion of these materials . . . will be dropped on the soil and . . . over generations there will be constant replenishment of the soil surface with steroidal substances of this kind. This in turn has its effect potentially on surface water-supply contamination and also potentially on the vegetable content of steroids in crops raised on such soil. . . . I think that we are now actually setting up a steroid cycle in our environment, and we have to give very serious consideration to its implications for our subsequent development and growth and possibly reproductive functions."[51]

One of the research assistants in Hertz's laboratory was John McLachlan, who went on to lead investigations into the human-health effects of DES exposure. McLachlan's work soon became pivotal in the understanding of environmental estrogens. After McLachlan read Herbst's 1971 *New England Journal of Medicine* article, he realized that both DES and DDT were estrogenic. During his postdoctoral research at the National Institutes of Health, he began looking at the long-term effects of DES as an "'estrogenic surrogate' for DDT," focusing on DES as a good chemical model for investigating the effects of DDT. His experiments on DES during the 1970s led him to question the broader implications of releasing DES into the environment through livestock, as well as the potential implications of DES exposure during prenatal human development.[52]

After the November 1971 Fountain hearings on DES in livestock, the FDA responded by increasing the withdrawal period of the drug from two to seven days before the cattle were slaughtered, so that they would have a full week free of any DES intake in their feed. Edwards gave his assurance that this would take care of the residue problems. In response to growing pressure from Congress, the USDA had already increased its residue-testing program, sampling 6,000 cattle in 1971 instead of the previous year's 192 cattle. The USDA informed Congress in early 1972 that not a single animal had been found with detectable residues. But the agency staffers were lying, suppressing results from 10 cattle with extremely high DES residues. When senators challenged the figures, the agency claimed that staffers were waiting until they could corroborate the results by another method of analysis before reporting the data. But in fact, no other method of analysis existed.

Members of Congress were furious at this duplicity. Senator William Proxmire introduced a bill to ban DES in livestock feed, while Senator Edward Kennedy called hearings on the problem of DES residues in beef.[53] At the Kennedy hearings, Proxmire argued that the FDA had refused to act on DES because staff members' theories got in the way of their observations of women. He asked, "With a 7-year history of residues, why hasn't the FDA banned DES under the Delaney Clause? The answer they gave last winter was to insist that there is in theory no reason for there to be residues. . . . It is not enough for the FDA to assert that their procedure should prevent all residues. It cannot comfort Americans to know that their meat is uncontaminated in theory, while it, in fact, may contain carcinogenic residues."[54] Proxmire identified a critical issue: the FDA's reliance on theoretical models had gotten in the way of empirical observation.

Commissioner Edwards and Simmons testified, as they had in the earlier hearings, that there was "no evidence" DES was a human carcinogen, in direct contradiction to every other scientist who testified at the hearing.[55] What was new about the Kennedy hearings was that the FDA's own research staff had spent the winter undermining the commissioner's claims. As FDA memos reveal, Leo Friedman, the director of the Division of Toxicology, had found new evidence that DES was significantly more toxic and "persists in man longer than estradiol."[56] Friedman's studies also refuted the second element of the FDA and industry argument: that DES residues were trivial compared to the amounts of estrogen naturally occurring in meat or in humans. Friedman found that the amount of estrogen in a single half-pound (227 gram) serving of liver at 2 parts per billion residue would contain nearly twice as much estrogen as a cow needs to go into heat. If that level of DES could put a heifer into heat, what might it do to a woman or a man? Moreover, Friedman argued, DES residues in meat were significant when compared to naturally circulating levels of estradiol as well. These residues could actually increase normal circulating estrogen levels in a postmenopausal woman from 300 percent to 2,200 percent.[57] New studies revealed that even incredibly tiny amounts of DES caused cancer in mice, which meant that no residue-testing program existed that could guarantee that the drug was safe.[58]

The FDA commissioner had argued that abundant scientific research proved that a seven-day withdrawal period ensured that no residues remained in the meat. Friedman's memo, however, undermined this argument. In fact, as Friedman's colleague, the biostatistician Nathan Mantel, wrote in another memo, Edwards was basing his claims on a single study with a single cow. Even worse, that cow *did* have DES residues in its meat, even after a ten-day withdrawal period.[59] Proxmire acquired copies of these memos and included them in his commentary, making certain that no one in the higher levels of the FDA could bury the evidence of agency staff.[60]

After the hearings, an editorial in the *Washington Post* castigated the FDA and USDA for incompetence, calling for "an independent consumer protection agency that would be empowered to act against such unacceptable bureaucratic behavior." With Proxmire's legislation and environmental lawsuits threatening, the FDA was forced to act. Yet the agency ruled only against the use of DES during pregnancy, not against its use in cattle. The day before the Kennedy hearings began, eight full months after the FDA had received Herbst's study, the FDA finally issued a warning that DES was contraindicated for use during pregnancy. Diane Dutton notes that doctors wrote more than sixty thousand DES prescriptions to pregnant woman during the months the FDA failed to act. Even after the tardy FDA warning, doctors could continue to prescribe the drug to pregnant women if they wished to do so because a warning had no legal power. Lilly's records indicate that sales of DES actually increased by 4 percent between 1971 and 1972.[61]

Edwards continued to resist all pressure to ban DES for cattle, arguing that illegal residues were the fault of bad farmers, not the drug itself. In 1972, however, new technologies were developed for residue testing, and these tests revealed that illegal DES residues would be present even when farmers and feeders followed the law precisely. No matter how careful a farmer was, DES got into the meat, and that was not legal. The FDA was finally forced to follow the law, and on August 4, 1972, Edwards announced that because of the Delaney Clause, DES had to be phased out of feed additives. Edwards did not hide his disdain for the clause, announcing that "under the law, there is no alternative but to withdraw approval of the drug, even though there is no known public health hazard resulting from its use."[62] Feed companies immediately sued the agency over the ban, and a federal court of appeals overturned it, ruling that before the chemical could be banned, the agency had to first perform a quantitative risk assessment that weighed the economic benefits of DES against the economic costs of cancer.

Even while DES use in cattle was being banned, its use to stunt the growth of tall girls was expanding. In 1946, a research abstract had suggested that DES could help limit the height of girls who were "becoming alarmed and unhappy about the extremes to which their exuberant, albeit normal growth was carrying them." The abstract received little attention until 1956, when the endocrinologist M. A. Goldzieher began promoting DES to treat what he called "excessive growth in the adolescent female." His publications on the "problem of tall girls" caught the attention of fellow physicians. Soon doctors were treating young American girls with diethylstilbestrol to stunt their growth. By 1977, long after DES use had been shown to cause cancer in women, fully half of surveyed pediatricians reported that they had treated tall girls with DES and related estrogens to keep them from getting taller than a feminine woman "ought" to be.[63]

The physicians Joyce Lee and Joel Howell describe the chilling ways that social beliefs about femininity shaped this emerging medical therapy. By the late 1960s, doctors had agreed that tallness was bad for girls, because a tall girl might be doomed to a life bereft of love and children. What man could desire a woman taller than himself? DES seemed to be the answer, even after reports of its link with vaginal cancer emerged in 1971. As Lee and Howell point out, "Idealized gender relations may be as important as scientific studies in determining what we will do as practicing clinicians," a lesson that applied to DES treatment for menopause and pregnancy as well.[64]

Calculations of risks and benefits had long been at the heart of debates over DES, and perceptions of those relative risks and benefits are always colored by cultural beliefs. In the 1970s, activists and consumers were outraged that health risks were being weighed against benefits to corporate profits, rather than benefits to patients' health. All drugs, new or old, bring with them some risk. Patients and consumers are willing to assume a degree of risk from a drug when the risk of *not* taking the drug is greater. While chemotherapy can cause terrible harm, many patients choose to accept its risks because those of forgoing chemotherapy are greater. The individual who bears the risk is also the individual who gets the benefit of the chemotherapy, a critical factor that was not true of most DES uses.

As early as 1939, the FDA had realized that the risk calculus for DES would be troubling. In trying to formulate a workable precautionary principle, FDA staff had recognized that if a woman chose to use DES to treat symptoms of menopause, the drug might improve the quality of her life, but it was not a lifesaving drug, so its risk calculations were different from those for a lifesaving antibiotic. The FDA had been reluctant to approve other quality-of-life drugs (such as topical estrogen creams) that threatened to cause liver damage and other serious harm, particularly when the patient might use the drug for years. Theoretically, when a woman took DES for menopausal symptoms, she was making the choice to assume the risk in return for benefits to herself. The fact that doctors and drug companies hid the true risks from patients muddled the ethical issues of risk, but the principle remained important to the FDA: an individual might choose to assume a risk in return for direct benefits.

The use of DES during pregnancy led to different risk calculations. The child would bear the potential risk, but the mother was making the choice to take the drug. The risk was one step deferred, although no one was certain what those risks might be. The benefits were equally uncertain. Many doctors treating diabetic patients were willing to take on these uncertain risks because diabetic pregnancies usually ended in failure, and the doctors and patients tended to assume that without the drug the fetus would probably die. Yet when the drug was allowed to be marketed to all women with normal pregnancies, not just diabetic pregnancies, the ethics of those risk calculations became particularly problematic.

In the case of DES in livestock, the risks and benefits became even more disassociated. The consumer was assuming the risk of DES residues but receiving none of the benefits, which went to the industry. Like the fetus, the consumers had no choice in the matter; they were never informed that they were consuming DES or DDT, much less told what the risks of those hormones might be.

Risk calculations within the federal regulatory agencies were intended to be quantitative and objective, yet they were beset by cultural assumptions about women. When members of Congress urged the FDA commissioner to issue public warnings alerting women to the risks of DES exposure during pregnancy, Edwards refused, explaining that he had to be careful "not to create an emotional crisis on the part of American women." Only doctors, not female patients, could comprehend risk, he argued; women were far too emotional to be informed that they were being exposed to a carcinogen. The medical establishment concurred with the view that women were unable to understand risk. The *Journal of the American Medical Association* editorialized that it would be inadvisable for the medical profession to inform women who took DES of the risks they had been exposed to; the physician should control the information. The editor explained, "The fact that a risk exists should be known to the physician and guide him to act in a careful, responsible fashion."[65]

Drug companies were no more eager than physicians or the federal agencies to inform women of the risks of DES exposure. While drug companies spent billions of dollars on issues concerning DES and cancer, that money largely went to defense against litigation, not to information or health care for the patients who had been exposed. Women found it nearly impossible to access their own medical records to determine whether they or their mothers had taken DES. Not until the women's movement became involved did DES-exposed individuals learn of their exposure and treatment options. Mothers and children exposed to DES joined with consumer activists to testify at congressional hearings and formed a national network called DES Action to direct research on DES exposure. The conflicts over DES mark a critical moment in the shift from technocratic expertise to political and consumer advocacy over chemicals. As the historian Rima Apple writes, DES debates highlighted troubling questions for regulatory agencies and consumers alike: "How can we determine the long-term effects of minute doses of potentially carcinogenic substances? What is a functional and acceptable assay method? What constitutes evidence of harm or safety? When science and technology did not provide simple, unequivocal answers to these questions, nonscientists entered the arena with their own interpretations of the evidence."[66]

There is no great mystery about why Lilly and other companies insisted that DES was safe long after evidence had accumulated showing that it was not. The drug brought enormous profits, and the cost of fighting bans on it in court was trivial compared to those profits. But why did the regulatory agencies, particularly the FDA, resist obeying the law for so long? Why invest so much in defending the industries the agency was supposed to be regulating, rather than in defending consumers? For some FDA employees, the reasons were economic. Many higher-level political appointees at the FDA had intimate ties to the industries they were supposed to be regulating. Just as a pharmaceutical company hired Theodore Klumpp from his FDA staff position immediately after he guided the approval of DES through the agency in 1940, so too did other regulators leave their agencies to work for the industry. Civil service pay was trivial compared with what an industry consultant could earn.

Most regulators were not driven by greed, however; they sincerely wanted to protect the public good. Yet it is never easy to define public good, especially when the regulators are technocrats with closer cultural ties to experts from industry than to consumer and environmental advocates. Moreover, few people find it easy to admit an error. In 1972 the journalist Nicholas Wade (now a science writer at the *New York Times*) puzzled over this question, concluding that "the defense of the carcinogenic food additive is a self-sustaining activity, from which the FDA can withdraw only at the price of admitting that the critics were right all along."[67] The longer the FDA defended DES, the harder it became for the agency to reverse course.

During the years that the regulatory agencies failed to ban DES in livestock, the chemical saturated environments and bodies. The drug moved inward into women's bodies, making its way into organs and cells, even transforming the genetic code, with effects that cross generations. The drug moved outward into broader ecosystems as well, when the metabolic byproducts of DES were excreted by humans and feedlot cattle. We are still struggling to comprehend the risks those residues pose to the people and wildlife exposed to them.

CHAPTER 7

Assessing New Risks

During the 1970s research into environmental pollutants flourished, and federal agencies made renewed efforts to regulate their risks. "Toxic torts," a type of personal-injury lawsuit in which a plaintiff claims that exposure to a chemical caused injury or disease, were becoming increasingly common, drawing consumer attention to health problems associated with chemical exposure. Yet for all the successes of the environmental movement, the government largely failed to control the growing risks from many of the new synthetic chemicals. Industrial food production increased its reliance on chemical inputs, and when courts upheld the ban on DES in livestock in 1976, other steroid hormones took its place. Similarly, after the Environmental Protection Agency (EPA) banned DDT for agricultural uses in 1972, other pesticides with equally toxic actions took its place.[1]

As environmental protection became a popular cause, lobbying associations such as the Manufacturing Chemists' Association and the National Agricultural Chemicals Association emerged as powerful forces working to counteract environmental restrictions and toxic torts. Companies and their lobbying associations did their best to influence government regulations and convince juries that toxic torts were unfounded, and when scientists and consumers attempted to push enforcement of stronger regulations, the industry fought back with lawsuits.[2] The result was a regulatory and judicial logjam: by the end of the 1970s, court cases and prolonged disputes had kept most toxic chemicals from regulation, with a few prominent exceptions, such as DDT. Then in 1980, Ronald Reagan won the presidential election, ushering in an era of intense hostility to regulation.

After Reagan took office, a technique called "quantitative risk assessment" came to dominate the oversight of most occupational and environmental chemicals in the United States. The assumption guiding quantitative risk assessment is that risk is an unavoidable fact of modern life, something to be managed rather than eliminated. Risk assessors estimate the size of a given risk posed by a given chemical, and risk managers decide whether that risk is acceptable. This process relies on estimates of how much exposure an average person can have to a particular toxic chemical without suffering significant harm. "Harm" is typically defined as cancer. Other endpoints, such as reproductive failure, are rarely taken into consideration. Risk assessors then manage the danger from toxic chemicals by permitting them to be used as long as they do not exceed a standard of contamination deemed to be an acceptable trade-off for economic gains. Who decides what trade-offs are acceptable to whom is deeply contentious, of course.[3]

Pressure from politicians, industry, the courts, and professional toxicologists all influenced the shift to quantitative risk assessment. In the 1970s, when the new Environmental Protection Agency and the Occupational Safety and Health Administration had attempted to use the Delaney Clause to restrict industrial carcinogens, industry had responded with both litigation and lobbying. One result was the Toxic Substances Control Act (1976), which forced agencies to balance the harm of potential deaths against potential economic benefits. Industry then took the EPA to court, forcing it to provide a quantitative rationale for regulatory decisions. Industry lawyers argued that qualitative judgments of potential harm to people from chemicals known to cause cancer in animals were "unscientific" and "unbalanced," for they failed to account for benefits from economic growth. The U.S. Supreme Court ruled in 1980 that qualitative methods such as those implicit in the Delaney Clause, which assumed that any dose of an animal carcinogen was probably also a human carcinogen, were insufficient to support regulation. Quantitative risk assessment, not just qualitative judgments of risk, needed to form the basis of regulatory action.[4]

Professional toxicologists helped promote the shift to quantitative risk assessment, in part because they were frustrated by the Delaney Clause, which they felt devalued their professional expertise. One toxicologist, Philippe Shubik, complained, "Unlike other manifestations of toxicity encountered in routine tests, 'carcinogenicity' has been selected for special treatment and the toxicologist in the United States has been told by the Congress how to use his findings. The Delaney Clause of the Food Additives legislation deprives the toxicologist of his usual prerogative to determine a rational approach to the control of a toxic manifestation of a substance. . . . All this is very sad since, although caution is necessary in introducing new chemicals, particularly into our food supply, this should not require a gross departure from the general principles of toxicology."[5]

New technologies for detecting the potential of a chemical to cause DNA mutations also shaped the shift toward quantitative risk assessment. The growing focus on DNA mutations as a source of cancer meant that research on environmental links to cancer became unfashionable and soon unfundable. If new technologies could identify the chemicals that caused the mutations, there would be no need for the Delaney Clause, toxicologists argued. Quantitative risk assessors could manage the risk scientifically. The focus on cancer, and the belief that the causes of cancer could be identified, known, and managed with simple assays for mutagenesis, all influenced the shift away from the precautionary principle of the Delaney Clause.[6]

Some environmental groups supported the shift to quantitative risk assessment, hoping to ease the regulatory logjam. Yet most were dismayed, arguing that risk assessment creates more problems than it solves. For endocrine-disrupting chemicals, the problems of assessment are particularly acute because the risk protocols were not designed for the biological mechanisms of endocrine systems.

The key assumption of risk assessment is that human bodies can accommodate some degree of chemical exposure as long as the exposure is below a threshold of toxicity.[7] Although this "dose makes the poison" concept may be true for drugs such as aspirin, it is rarely true for endocrine disruptors. At low concentrations, hormones normally stimulate receptors, but at high concentrations hormones can saturate receptors, thus inhibiting their pathways. Low doses of endocrine disruptors such as DDT or DES might produce adverse impacts, even though higher doses might not. But the idea that a substance can have more powerful effects at low doses than at high doses fundamentally challenges traditional approaches to toxicology and with them quantitative risk assessment models.[8]

Quantitative risk assessments have focused on the potential of chemicals to cause cancer by directly damaging DNA and thus leading to genetic mutations. Endocrine disruptors, however, often do not alter genes. Instead, they change the way genes are expressed, an outcome which is not captured by risk assessment. Risk assessments typically measure effects on adults, but with endocrine disruptors, fetal effects are more problematic. A tiny exposure of the fetus may have effects that are not obvious at birth, yet decades later that fetal exposure may result in reproductive failure.[9]

Because most endocrine disruptors are persistent, toxic, and accumulate in the environment, they challenge the fundamental assumptions of quantitative risk assessment. Like DES, modern synthetic chemicals have pervaded global environments, crossing international boundaries to contaminate landscapes, bodies of water, wildlife, and people far from the initial source of production and consumption. These chemicals have transformed the internal ecosystems of human, livestock, and wildlife bodies, interconnecting our bodies with our environments in increasingly complex ways. Toxic residues complicate not only spatial boundaries but also temporal distinctions, for their effects persist long after exposure.[10] An examination of the use of steroids in modern livestock and the chemicals in modern plastics will illuminate the ways that endocrine-disrupting chemicals complicate quantitative risk assessments.

Steroids in Beef

Problems with quantitative risk assessment and its approach to regulation can be seen in the continuing use of steroid hormones in livestock. The ban on DES in beef cattle was finally upheld in 1979, but six other steroids replaced DES. Steroids are now used in 90 percent of American beef cattle, while bovine-growth hormones (which are proteins, not steroids) are used in most dairy cows.[11] Currently three naturally occurring steroids (estradiol, testosterone, and progesterone) and three synthetic steroids (zeranol, a synthetic estrogen; trenbolone acetate, a synthetic androgen; and melengestrol, a synthetic progestin) can be used in beef cattle, often in combination. The FDA acknowledges that when livestock are treated with steroids, measurable residues of known carcinogens remain in meat intended for human consumption. Under the Delaney Clause, these residues would have been outlawed. But because the clause was repealed in

1996, it is now legal to include these known carcinogens in food if the hormone levels do not exceed what the FDA calls "acceptable daily intakes" for each of these drugs.

Acceptable daily intakes are derived from quantitative risk assessments that rely on traditional toxicological testing. The tests determine how much of the chemical is required to lead to DNA mutations in adult laboratory animals. While mutations are significant, the tests ignore the fact that mutations may not matter to fetal health as much as epigenetic modifications of DNA, which change the programming of genes during differentiation and which typically occur at lower levels of exposure. Nor do the risk assessments take into account the fact that fetuses and young children respond to hormones at very low levels, often much lower than those that adults do. In adults, exposure to hormone residues might lead to transient physiological effects that vanish when the exposure ends. But in fetuses, exposure to low doses can have lifelong effects.[12]

Both the FDA and USDA insist that their quantitative risk assessments show that steroid hormones in beef are safe for human consumption, even though the European Union and many other countries worldwide refuse to allow them. The FDA and USDA arguments in favor of the hormones rely not on objective, scientific precision, as the agencies imply. Rather, they depend on two arguments familiar from the 1930s: the "natural" argument and the "technological-control" argument. The U.S. regulatory agencies insist that because several of the steroids occur naturally in cattle, they must be safe since synthetic steroid residues are only a small fraction of the levels of hormones occurring normally. Although industry made this claim as early as the 1930s, the FDA itself frequently refuted it. As Roy Hertz testified in 1971, just because something is natural does not mean that it is safe. If a natural carcinogen exists and already contributes to cancer incidence, increasing the levels of the carcinogen by adding it to the diet is unlikely to benefit public health.[13]

In addition, the FDA currently claims that steroids in beef are safe because technology allows experts precise control over synthetic hormone levels. A similar claim was made in 1944 by Wick and Fry, however, and the FDA rejected the argument. FDA staff continued to reject the argument into the 1970s. The history of DES in livestock shows that models which rely on technological controls for safety may mean little in actual practice. Hormones from implants continue to make their way into the human food supply, and when agencies test for residues, they consistently find them. In one study of thirty-two feedlots conducted during the 1990s, fully half the implants were illegally present in edible muscle tissue, greatly increasing consumer exposure to hormones.[14] Nevertheless, the official FDA position relies on the idea of model conditions in which technological control is possible.

Steroids in beef have become the flashpoint for a trade war between the United States and the European Union, a conflict that centers on perceptions of risk. After DES appeared in Italian baby food in 1980, infants of both sexes who had consumed the tainted food developed breasts, and some infant girls reportedly began menstruating. The European Union responded by outlawing the use of all growth-promoting hormones in cattle, and in 1989 it banned imports of U.S. beef treated with hormones. The United States objected, taking the case to the World Trade Organization, which ruled against the ban in 1997, arguing that the European Union had not conducted a formal quantitative risk assessment establishing specific harm to individuals. The European Union appealed the decision, arguing that quantitative risk assessments were inadequate and that in the face of scientific uncertainty the precautionary principle should prevail.[15] After losing the appeal, the European Union agreed in 1999 to conduct a formal quantitative risk assessment, which eventually determined that no safe exposure level could be determined for children.[16] In addition, the risk-assessment panel pointed out that the FDA and USDA had no scientific basis for their assurances that the meat was safe because "none of the approximately 130 million U.S. livestock slaughtered annually are tested for residues of cancer-causing and gene-damaging estradiol or any related sex hormones."[17]

When the *Wall Street Journal* reported in 2000 that DES had been detected in U.S. beef shipped to Switzerland, anger in Europe was intense. American consumers, alerted by the media about the European Union bans and the DES detected in U.S. beef exported to Switzerland, also became concerned. The FDA responded by reassuring American consumers that no DES had been found by any regulatory authorities on any U.S. farm for the past sixteen years. But, as critics pointed out, the agency had never found DES because it had never looked for it. As the *Los Angeles Times* had earlier reported, "A USDA spokesperson confirmed that the agency has never tested for the natural hormones and has done only sporadic testing of the synthetic ones, almost all of it before 1990."[18] In 2009, the trade war is still unresolved. The U.S. government continues to claim that scientific risk assessment and monitoring will ensure the safety of consumers exposed to hormonally treated cattle, while the European Union continues to forbid imports of U.S. beef treated with hormones.

As the trade war illustrates, the health risks of exposure to hormone residues in livestock remain controversial. Hormone residues are usually present at very low levels, but exposures are chronic, stretching from conception until death. Current epidemiological tests cannot determine safety in such situations, and historical analogies may offer better guides to the potential risks of low-level, chronic exposure to endocrine disruptors in food. DES residues in beef offer such a historical analogy.[19]

Essentially everyone in the United States was exposed to DES, so it is impossible to find a control group of unexposed people. But researchers have tracked disease rates in groups of people whose mothers ate different amounts of beef during pregnancy. Sons of women who ate more beef while pregnant have lower sperm quality and higher rates of fertility problems than sons of women who ate less beef. Because the women were pregnant during the time that essentially all U.S. beef cattle were treated with DES and other steroids, women who ate more beef were probably exposed to more steroid residues. In laboratory experiments DES residues have been shown to affect testicular development, so it is reasonable to suspect that DES residues in beef may have affected human testicular development in the womb. These findings are also consistent with experiments on mink colonies that link residues of DES to reproductive failures. However, the possibility remains that other substances in beef, such as heterocyclic amines produced during cooking or pesticides in the fat, could have contributed to the fertility problems of these men.[20]

Other studies suggest that very low levels of hormonal residues in meat may have affected female reproductive health. No one has yet published studies on the daughters of women who ate beef while pregnant in the 1960s and 1970s to see whether maternal beef consumption correlates with the development of fibroids, infertility, and other DES-related issues in adulthood. However, a study of more than ninety thousand American women found that those who ate more beef were more likely to develop estrogen-responsive breast cancer. This finding is consistent with another startling finding published in 2006 showing a decline in breast cancer following a drop in women's use of hormone-replacement therapy. Like beef consumption, hormone-replacement therapy also exposes women to very low doses of synthetic estrogens. After reports in 2002 from the Women's Health Initiative suggested that hormone-replacement therapy increased the risk of heart problems, its use declined sharply. Rates of breast cancer soon decreased as well, suggesting that low-dose exposure to estrogens was increasing breast cancer incidence.[21]

Children's consumption of meat tainted with hormonal residues may have contributed to breast growth and early onset of puberty in some cases. Eating DES-tainted meat appeared to correlate with more than five hundred cases of early puberty in Puerto Rico, although those findings are contentious. In Italy, DES residues in school luncheons (as well as in baby food) were suspected as the cause of breast growth in young girls and boys, but again, conclusive proof was lacking.[22]

The effects of hormone residues on wildlife are excluded from quantitative risk assessments, but those effects may be profound. When a bit of manure excreted by a cow finds its way into a river, the chemicals she was fed end up in the river as well. A fish swims in the chemical soup, downstream from where the cow had grazed. When a pregnant woman eats that fish, the pollutants may affect her son's development, potentially undermining his future fertility.[23]

When biologists sample the health of a river, they often find aquatic systems transformed by endocrine disruptors. During 1999 and 2000, the Toxic Substances Hydrology Program of the U.S. Geological Survey collected and analyzed water samples from 139 streams in thirty states. Researchers measured concentrations of 95 wastewater-related organic chemicals in water. Eighty percent of sampled streams contained at least one of the chemicals, and 75 percent of the streams contained chemical mixtures. Streams that flow downstream from feedlots were particularly high in chemicals, a finding of great concern for fish health. Because fish gills concentrate whatever contaminants exist in water, fish tend to be particularly vulnerable to aquatic pollution. In addition, because fish are bathed in water from birth until death, they serve as excellent monitors for the health of stream systems.[24]

In the 1990s, Ana Soto of Tufts University Medical School examined hormones moving off feedlots in Nebraska, comparing the hormonal activity of water sites downstream of feedlots with that of water collected

Feedlot runoff has proven to be contaminated with hormones, as Roy Hertz predicted in the 1950s (photograph courtesy of USDA Natural Resources Conservation Service)

upstream. She found that concentrations of estrogenic pollutants at two of the downstream sites were almost double those at the upstream site. Water from all three downstream sites was significantly more androgenic than that at upstream sites. Steroids from these Nebraska feedlots might have been disrupting hormones in fathead minnows: males downstream of the feedlots had smaller testes and produced less testosterone than males upstream.[25]

In 2002 sites downstream of cattle feedlots were found to have significant levels of hormonally active compounds and fish with altered sexual development. In 2004 scientists tested twenty sites downstream of feedlots for hormones, and more than half the sites had detectable levels of estrogenic activity. Estrogenic chemicals from feedlots, both from added hormones and from the animal's own hormones, were present at concentrations high enough to harm reproductive development in fish.[26] Roy Hertz was prescient in his concern that treating livestock with synthetic hormones might be creating a steroidal cycle in the environment. We have indeed altered the sexual development of aquatic life with the hormones we have given livestock. What that means for human reproductive health is still unclear.

Hormones fed to livestock create cascading changes to hormone systems within the environment and within the consumer. The meat from animals treated with hormones may contain direct chemical residues. Indirectly, the meat of animals fed grains has a nutritional profile that is different from that of grass-fed animals, and those differences include complex hormonal transformations. Grain-fed cattle, for example, are deficient in omega-3 fatty acids, which modulate the steroids involved in metabolism and reproduction. Foods low in omega-3 fatty acids may increase a woman's risk of developing polycystic ovary syndrome, a hormonal imbalance that affects fertility.[27] Grain-fed animals also contain less of a particular fatty acid known as conjugated linoleic acid. Conjugated linoleic acid blocks estrogen signaling in human breast-cancer cells, and many epidemiological studies link consumption of linoleic acid with lower rates of breast cancer. So in addition to its direct effects on breast cancer and other hormonally related conditions, the use of steroids in livestock production may indirectly increase reproductive problems in humans. Yet none of these problems can be quantified by risk-assessment protocols, and so they are ignored when it comes time to regulate.[28]

Plastics and Regulation

Many modern plastics leach extremely low levels of endocrine-disrupting chemicals into the environment, levels so low that conventional risk assessments have minimized their danger. Yet because plastics have become ubiquitous in our daily environments, we are exposed to these chemicals from birth to death. When levels are low but exposure is chronic, quantitative risk assessments offer little protection against potential harm, as the difficulties of regulating plastics illustrate.

Plastics originated in the 1860s, when the British researcher Alexander Parkes synthesized a material known as Parkesine, which soon proved cheaper to produce than rubber. During World War II, engineers developed an array of new, lightweight plastic military materials ranging from weapons to food packaging. After the war, demand skyrocketed. DuPont, the chemical company that became a major plastics producer, famously promised Americans "Better living through chemistry." Dow, Monsanto, and Union Carbide joined DuPont in promoting plastics, launching advertising campaigns that depicted plastic as a modern miracle. The cultural historian Jeffery Meikle shows that the postwar growth of the chemical industry resulted from a number of interconnected economic, cultural, and political factors. A boom in petroleum production made plastic seem cheap, the media made plastic seem modern and hygienic, while the widespread faith in science and technology made plastic seem harmless. And finally, the growing power and influence of the chemical companies made objections to plastic's growing ubiquity seem pointless.[29]

Plastics had promised to do away with dirt, imperfection, and natural disorder. In 1941 two British chemists had imagined "a dweller in the 'Plastic Age'" named Plastic Man, who would be born "into a world of color and bright shining surfaces, where childish hands find nothing to break, no sharp edges or corners to cut or graze, no crevices to harbour dirt or germs." He would be "surrounded on every side by this tough, safe, clean material which human thought has created." As Meikle writes, quoting the chemists, "The world of Plastic Man would be 'brighter and cleaner' than any previously known, 'a world free from moth and rust and full of color.' . . . The new civilization—'built to order' by industrial chemists and technocrats—would embody 'the perfect expression of the spirit of scientific control.'"[30]

Because plastics are cheap, light, and durable, they have become an inseparable part of our everyday environments. But for those same reasons, plastics have also insinuated themselves into some of the most remote and wild places on the planet, as well as into our most intimate bodily spaces. Donovan Hohn's provocative *Harper's* magazine essay on plastic ducks, "Moby-Duck," describes a garbage patch of plastic that swirls in the North Pacific ocean: "Approximately 800 miles west of California, where the wind speed fell below ten knots, drifts of garbage began to appear. The larger items that [the researcher Charlie] Moore and his crew retrieved from the water included polypropylene fishing nets, 'a drum of hazardous chemicals,' a volleyball 'half-covered in barnacles,' a cathode-ray television tube, and a gallon bleach bottle 'that was so brittle it crumbled in our hands.' Most of the debris that Moore found had already disintegrated. Caught in his trawling net was 'a rich broth of minute sea creatures mixed with hundreds of colored plastic fragments.'"[31]

In the early 1990s, I encountered a similar plastic stew when I was a zoology graduate student at the University of Washington. One of my projects involved dissecting dead albatrosses trapped as bycatch in squid drift nets. Fishing fleets using these drift nets trawled the North Pacific ocean, as remote from industrial civilization as any place on earth, yet much to our surprise, nearly every albatross we opened up contained plastic fragments in its stomach. In addition, the albatrosses had incomplete molt patterns (reducing their ability to breed) and heavy parasite loads in their esophagi indicative of reduced immunity. We published a series of papers theorizing about the evolutionary history of these deleterious traits, but we never imagined that they might also be related to the chemicals leaching out of the plastics in their guts.[32]

In 1987, Ana Soto and Carlos Sonnenschein at Tufts University stumbled across the surprising fact that certain plastics might be leaching chemicals with estrogenic effects. The two had been examining the ways estrogens make breast-cancer cells multiply. For their research, they used a line of breast-cancer cells that grow only in the presence of estrogens. But they suddenly discovered that breast-cancer cell cultures had started growing and dividing on their own, even before the experiments had started, when nobody had added anything to them. Soto assumed that someone in the lab had been careless and contaminated the clean cells or had mistakenly added estrogen to the wrong cell lines. But eventually she realized no one in the lab had made a mistake. The problem lay in the new plastic tubes: something from these tubes was leaching into the cultures and stimulating the growth of breast-cancer cells. The manufacturer had recently changed the formulation of the tubes, and those sterile tubes were now leaching a chemical that acted like an estrogen. This astonished Soto and Sonnenschein because they knew of no reports concerning estrogens leaching out of plastics. The new chemical turned out to be something known as nonylphenol. Nonylphenols are widely used in industrial and domestic products such as paints, detergents, oils, toiletries, and agrochemicals. They are part of a class of related chemical compounds called alkylphenols, many of which are weakly estrogenic and make breast-cancer cells multiply in lab cultures. In Britain these chemicals were found in rivers and lakes at concentrations of 50 micrograms per liter, higher than needed to induce cancer-cell responses in the lab.[33]

Another common chemical found in plastics, bisphenol A, illustrates some of the challenges that plastics, with their estrogenic chemicals, pose to federal regulators. Bisphenol A has fascinating parallels with diethylstilbestrol, both in its estrogenic effects on living creatures and in its political history. Between 1980 and 2000, the production of bisphenol A grew sixfold. Almost three billion kilograms of the chemical are now produced each year, and sales generate billions of dollars annually. Bisphenol A has become extremely common in American households. Those convenient Nalgene bottles, popular with students and hikers, canned food, bicycle helmets, laptop computers, car parts, coffeemakers, baby bottles, dental sealants, water filters, and even some paper: all contain (or recently contained) bisphenol A.

The public health historian Sarah Vogel argues that whenever scientists have published research questioning the safety of bisphenol A, the industry has countered with massive public relations campaigns. The journalist Elizabeth Grossman notes that the American Chemistry Council, an industry trade association, "tells us that bisphenol A makes our lives 'healthier and safer, each and every day.'" Many scientists, consumers, and environmentalists disagree, as research accumulates that links this ubiquitous and estrogenic chemical with infertility, children's reproductive health, obesity, breast and prostate cancer, and neurological disorders.[34]

Synthesized in 1891 by the Russian scientist A. P. Dianin, bisphenol A has two phenol rings joined together by a carbon bond. Bisphenol A, in fact, was one of the chemicals that Dodds and his colleagues noted "excited oestrus" in his lab animals when they were searching for cheap, synthetic estrogens. In 1936 the researchers announced their discovery that bisphenol A was a reasonably potent estrogen, though not as powerful as the natural estrogens. Endocrinologists expressed interest in the chemical's potential as a commercial estrogen, but this interest waned in 1938 when Dodds synthesized diethylstilbestrol, which proved far more estrogenic.[35]

Bisphenol A received little attention for the next fifteen years until chemists discovered they could polymerize it, forming long chains that would become a key constituent of hard, clear polycarbonate. Polymerization appeared to promise an inexpensive, stable plastic with many uses. Unfortunately, however, the bond that links the monomers is not particularly stable in water or food. When the bond decays, bisphenol A leaches out of the plastics that contain it, often making its way into our bodies.[36]

Because most food cans are lined with epoxy resin containing bisphenol A, canned food is directly exposed to the chemical. Bisphenol A is also released from polycarbonate containers, even at room temperature. Infants fed formula from polycarbonate bottles can consume bisphenol A in quantities of up to 13 micrograms per kilogram of body weight each day, higher than those causing significant effects in animal studies. Largely through food and drink containers (although dental sealants are also a source of exposure for children), bisphenol A has made its way into nearly all Americans. Ninety-five percent of adults sampled in 1988–1994 and 93 percent of children and adults sampled in 2003–2004 had the chemical in their urine. The levels ranged from 33 to 80 nanograms (one billionth of a gram) per kilogram of body weight. Levels found in humans are about a thousand times lower than the 50 micrograms (one-millionth of a gram) per kilogram of body weight per day that the FDA considers safe for adults, so both the FDA and industry have assumed that little cause for concern exists. Yet whether those levels are safe for developing infants is increasingly doubtful.[37] Plastics in contact with food are allowed to leach bisphenol A as long as the levels remain lower than the FDA standard. Yet the safety standard was developed from a single high-dose study done in the 1970s, and only two other industry-funded studies uphold the standard; meanwhile, the FDA has ignored hundreds of studies with contradictory findings.[38]

The first hint that bisphenol A might affect development came when a group of researchers led by Patricia Hunt had an experience very similar to Soto's misadventures with nonylphenol contamination. In 1998, Hunt and her colleagues were experimenting on the chromosomal changes that occur in egg cells from aging mice. The controls suddenly began showing numerous chromosomal abnormalities. Hunt traced the problem to contamination from bisphenol A released from degrading plastic in the animal cages. A year later Japanese researchers demonstrated that the placenta does not act as a barrier to bisphenol A, just as it does not block DES from reaching the fetus. Maternally ingested bisphenol A reached maximum concentrations in the fetuses of lab rats in only twenty minutes.[39]

Numerous laboratory animal studies have shown that bisphenol A can alter the behavior of more than two hundred genes, and these genes influence how cells multiply, how stem cells become more specialized, how metabolism is regulated, and how the brain develops. Many of these experiments find effects at very low doses, similar to doses absorbed by people in their ordinary lives. Nevertheless, just as with diethylstilbestrol, the industry insists that low-dose findings from experimental animal studies mean little for human health. Because no data convincingly prove harm to humans, the industry insists that the chemical must be safe at low levels. Environmentalists and other researchers point out that, just as with DES, a lack of evidence of harm does not imply safety. Because it is not legal or ethical to experiment with the chemical's effects on human fetuses, we cannot acquire direct experimental evidence of harm, even though an enormous, uncontrolled experiment is being performed on humans every day.

Research results vary tremendously depending on the source of funding, adding to the confusion. One striking observation is that government-funded studies have overwhelmingly found evidence that low doses of bisphenol A interfere with embryonic development in animals, whereas industry-funded studies have found far fewer effects. In 2005, the biologists Frederick vom Saal and Claude Hughes surveyed 115 published experimental studies on low-dose effects of bisphenol A. More than 90 percent of the government-funded studies showed significant effects at doses below the EPA's lowest adverse-effect level, whereas not a single industry study found any effect.[40]

Many researchers and environmentalists suspect that industry-funded studies are designed *not* to detect effects. The design of their controls is particularly contentious. Industry studies tend to use a strain of rat with little sensitivity to estrogen, and many of these studies also omit what are known as positive controls: that is, in addition to comparing the chemical with a negative control that is expected to cause no hormonal effect (water or oil, for example), researchers compare it with a positive control known to cause hormonal changes. If scientists want to test whether bisphenol A has estrogenic effects, their toxicological experiment should include three groups: a standard control group exposed to a substance known to cause no estrogenic effect; a positive control group exposed to a substance such as DES known to cause an estrogenic effect; and the experimental group exposed to bisphenol A. If the experimental group shows estrogenic effects while the standard control does not, researchers can conclude that bisphenol A is estrogenic. If the experimental group and the standard control groups both show no effect, and the positive control group treated with DES does show an estrogenic effect, then they can conclude that bisphenol

A is not estrogenic. But if all three groups, including the group treated with DES, show no estrogenic effect, researchers cannot conclude that bisphenol A has no effect. All they know is that the experimental design made it impossible to detect estrogenic effects, since the DES-treated positive control group did not show the expected estrogenic effects. Without a positive control group, scientists cannot determine whether a substance is safe. Perhaps it is, but equally the sample size may have been too small to detect significant effects, or the mice may have been from a strain that does not respond to estrogen.[41]

Researchers can perform endless experiments on lab animals and still disagree about whether the results are relevant for people. Industry trade groups argue that human experiments should be required for regulation, knowing that such experiments can never be performed. We must find another approach to determining the potential risk or safety of a chemical such as bisphenol A. By using a known fetal endocrine disruptor such as DES as a model, then comparing the effects of historic DES exposure to the effects of the chemical in question, researchers can evaluate potential risks from exposures without subjecting human populations to the chemical and waiting to see what happens.[42]

A series of experiments has shown that DES can indeed provide a useful model for evaluating the effects of low-dose human exposure to bisphenol A. Exposure to low doses of bisphenol A during pregnancy induces chromosomal abnormalities in the fetus during early stages of development, just as DES does.[43] With both chemicals, low-dose fetal exposure results in reproductive and metabolic changes that persist into adulthood. With both chemicals, many effects are invisible at birth and become apparent only at sexual maturity. With both chemicals, low-dose fetal exposures induce permanent, irreversible changes, even though adult exposures may have only transient effects. With both chemicals, exposing a pregnant female affects the mother, the developing fetus, and even a third generation (if there is one). Both chemicals induce epigenetic changes, altering patterns of DNA methylation and changing the activation and blocking of genes. Both chemicals can cause lasting changes in the development of the uterus that can pose reproductive problems at adulthood, including infertility and miscarriage. Both bisphenol A and DES induce fibroids in lab animals. Because women exposed to DES experienced many of the same effects that laboratory animals exposed to low doses of DES experienced, this suggests that low-dose bisphenol A exposure may also increase the risk of fibroids in women, as it does in laboratory animals. As with DES, bisphenol A exposure during pregnancy can induce changes in mammary-gland development that persist into adulthood.[44]

Biologically, both DES and bisphenol A are estrogenic, with fetal effects that become visible at maturity. Both alter epigenetic and metabolic processes, and both influence the signaling systems in the body. Both disrupt the normal toxicological paradigm, showing an array of low-dose effects that are often greater than, and sometimes completely different from, high-dose effects. Both DES and bisphenol A interrupt what the biologist Retha Newbold calls "conversations between cells, which are set up during prenatal development."[45] By interrupting these prenatal conversations, both chemicals have the potential to alter not just human health but also wildlife and broader ecosystems.

Because experimental animal studies of exposure to DES and bisphenol A find similar effects, many scientists infer that the results of human exposure to bisphenol A may be similar to those of human exposure to DES. A precautionary approach would therefore conclude that limiting human exposure to bisphenol A is advisable, given the effects that emerged from human exposure to DES. Ideally, regulatory agencies would learn from history, averting a repetition of the DES tragedy.

Industry advocates typically have countered this precautionary argument by pointing out that no human studies have demonstrated that bisphenol A exposures affect people. But in September 2008, the first major epidemiological study to examine the human health effects of bisphenol A exposure was published. This study showed that urine levels of bisphenol A correlated to increased rates of heart disease, diabetes, and liver-enzyme abnormalities. Correlation cannot prove causation, but because experimental animal studies also show adverse effects at low-doses, and because molecular studies identify the mechanisms of these responses, and because DES provides a model of possible human effects, it is reasonable to conclude that human exposure to low doses of bisphenol A poses significant risks.[46]

Politically, the cases of DES and bisphenol A have important parallels. DES was worth millions of dollars to the pharmaceutical industry, and when research that showed harm to laboratory animals threatened profits, the industry hired consultants to testify that animal studies had no significance for people. When research then showed that individuals were dying from DES, the industry argued in congressional hearings that those cases had no connection to DES residues in livestock. When epidemiological studies showed that DES residues in livestock might harm both wildlife and people, industry representatives such as Don Hines attempted to cast doubt on that research. And when government agencies finally stepped in and began attempting to regulate DES, the industry took the agencies to court, buying a few more years of profits by manipulating scientific uncertainty.

The same patterns are now unfolding for bisphenol A. Once again, the industry is employing lawsuits, delays, and attempts to discredit scientists in their fight for the chemical. Most important, scientific uncertainty is being exploited to block regulation. Magnifying discrepancies between different scientific studies and uncertainties about mechanisms of action allows the industry to delay action—a pattern with many precedents in public health history.

In a critique of corporate influence on public health, the epidemiologists and public health scholars David Michaels and Celeste Monforton write in the *American Journal of Public Health* that "scientific uncertainty is inevitable in designing disease prevention programs. . . . By magnifying and exploiting these uncertainties, polluters and manufacturers of dangerous products have been remarkably successful in delaying, often for decades, regulations and other measures designed to protect the health and safety of individuals and communities."[47] Just as the pharmaceutical industry delayed restrictions on DES, so it is delaying bisphenol A restrictions. As a tobacco industry document had proclaimed in 1969, "Doubt is our product since it is the best means of competing with the 'body of fact' that exists in the mind of the general public."[48] Three decades later, industry advocates for bisphenol A continued to cast doubt on researchers and their findings, manufacturing uncertainty to delay regulation.

In 1997, when Frederick vom Saal and his colleagues published research showing that low levels of bisphenol A could induce experimental responses similar to those shown by DES, the chemical industry attacked the researchers' credibility. As vom Saal recounts in an interview, "The moment we published something on bisphenol A, the chemical industry went out and hired a number of corporate laboratories to replicate our research. What was stunning about what they did . . . was they hired people who had no idea how to do the work." Soon a group of industry-funded studies were published that purported to invalidate vom Saal's work, and vom Saal was mocked in the "anti-junk-science" blogs and press.[49] When a study came out by Chandra Gupta, a biologist at the University of Pittsburgh, that replicated vom Saal's findings, a group of researchers from the Chemical Industry Institute of Toxicology published a commentary attacking her methods and conclusions.[50]

Attempts to regulate endocrine-disrupting chemicals face new challenges, however, that lack historical precedent. When DES research was being contested in the 1970s, the EPA was a young agency, and quantitative-risk-assessment protocols were still being developed. Regulatory agencies had not yet set up review panels of experts to read through the peer-reviewed research and pronounce judgment on conflicting studies. These review panels provide a powerful tool to help policymakers sort through conflicting scientific evidence, but they are also vulnerable to manipulation by interest groups from opposing sides.

After two review panels on the risks posed by bisphenol A reached different conclusions in 2008, the National Toxicology Program tried to reconcile the results.[51] On April 14, 2008, the first draft of the National Toxicology Program's brief on bisphenol A was released. When the media learned that this draft raised "the level of concern for the reproductive effects of bisphenol A at levels of human exposure," environmental groups responded with calls to regulate bisphenol A. After the *Today Show* aired a segment discussing the risks of bisphenol A for children, the American Chemistry Council responded by claiming that the *Today Show*'s decision to report on the risks of bisphenol A "reverses 40 years of industry science [and] strikes needless fear in millions."[52]

While environmentalists as well as industry advocates applaud review panel findings that support their perspective and denounce those that do not, the chemical industry has been accused of something more serious: attempting to unduly influence the makeup of bisphenol A review panels, thus distorting the entire process.[53] In 2008, Representatives John Dingell and Bart Supak began a congressional investigation on conflicts of interest in the federal agencies' reviews of toxic chemicals. Dingell wrote to EPA administrator Stephen Johnson, a political appointee of President George W. Bush's who had gained a reputation during his tenure at the EPA as a foe of environmental regulation. Dingell noted that "a number of EPA panels assessing the human health effects of toxic chemicals have included individuals alleged to have pecuniary interests in the chemical industry," and went on to list nine examples.[54] Dingell and Supak have also accused the FDA of "cherry picking research" to support its decisions about safe levels of bisphenol A. In particular, Dingell and Supak were troubled by the industry's use of product-defense firms such as the Weinberg Group. In their words, product-defense firms "manipulate public opinion related to certain chemicals," and manipulate the science as well.[55]

The Weinberg Group illustrates the tangled politics of scientific uncertainty, whereby industry can profit by casting doubt upon research. In the 1980s, the Weinberg Group had established its reputation with its work on behalf of the tobacco industry and manufacturers of Agent Orange and Teflon. In one letter to DuPont uncovered by the journalist David Roberts, a vice president of the Weinberg Group noted that the company "has helped numerous companies manage issues allegedly related to environmental exposures. Beginning with Agent Orange in 1983, we have successfully guided clients through myriad regulatory, litigation and public relations challenges posed by those whose agenda is to grossly over regulate, extract settlements from, or otherwise damage the chemical manufacturing industry." In 2003, DuPont was faced with possible litigation over a chemical in Teflon that had been linked to reproductive damage. The Weinberg letter described the many ways the firm could help DuPont avoid paying damages, for example "by analyzing existing data, and/or constructing a study to establish" that the chemical was safe and "offers real health benefits." The letter promised to "harness, focus, and involve the scientific and intellectual capital of our company with one goal in mind—creating the outcome our client desires." One way to create the desired outcome would be to discredit not just individual studies but the entire scientific discipline of epidemiology. The letter offered to "provide the strategy to illustrate how epidemiological association has little or nothing to do with individual causation," thus making it impossible for plaintiffs to win in court. In addition, the letter authors promised to recruit scientific experts "so as to develop a premium expert panel and concurrently conflict out experts from consulting with plaintiffs." In other words, experts who worked for DuPont through the Weinberg Group would have been unable to testify for plaintiffs in court. As Dingell and Stupak commented: "The tactics apparently employed by the Weinberg Group raise serious questions about whether science is for sale at these consulting groups, and the effect this faulty science might have on the public health."[56] As the continuing bisphenol A controversies illustrate, industry can use quantitative risk assessments to delay regulation because no one can ever deny that more research might help clarify risks.

Throughout 2008 and early 2009, U.S. federal agencies continued to delay regulation of bisphenol A, even though Canadian agencies responded to the new research results with restrictions on the chemical. On April 18, 2008, the Canadian health minister announced a ban on the import, sale, and advertising of polycarbonate baby bottles containing bisphenol A. As the Environment Canada Agency Web site announced, Canadian risk assessments had judged bisphenol A to be a "'toxic chemical' requiring aggressive action to limit human and environmental exposures." Even though U.S. risk assessments disagreed with the Canadian findings, a consumer tipping point over bisphenol A emerged during 2008. Walmart announced that it would cease sales in Canadian stores of all food containers, water and baby bottles, sippy cups, and pacifiers containing bisphenol A, and consumer pressure persuaded the company to extend the same restrictions to U.S. stores in 2009. Nalgene announced that it would stop using the chemical in some of its bottles. The online retailer Amazon.com began tagging products as "bpa-free," and virtual communities sprang up to discuss ways to create a bisphenol A–free life. Senator Charles E. Schumer (D-N.Y.) announced that he was going to introduce legislation to ban the use of bisphenol A in all children's products and "food contact" consumer products, including water bottles and food containers.[57]

The political controversies swirling around bisphenol A assessments reveal a key flaw at the heart of quantitative risk assessment. Regulators had hoped to construct an objective framework for assessing risk that was independent of culture, politics, and economics. But clearly, the FDA decisions about beef steroids and bisphenol A fail to demonstrate such independence. Risk assessment, in essence, has meant that society hands over regulatory decision making to what the environmental scientist Ellen Silbergeld calls "a technological elite, whose members alone can understand the increasingly arcane basis and data analysis of quantitative risk assessment."[58] In trying to sidestep politics with quantitative tools, regulators have found themselves caught in a tangle of competing voices and manipulated by increasingly powerful lobbying groups. We need to devise a more effective process that accepts the presence of continuing uncertainty.

CHAPTER 8

Sexual Development and a New Ecology of Health

In 2006 headlines proclaimed that bisphenol A made female mice act like male mice. Just as diethylstilbestrol had seemed to change gender norms, so did bisphenol A. "Boyish Brains" cried the headline of one popular science article: "Exposure to the main ingredient of polycarbonate plastics can modify brain formation in female mouse fetuses and make the lab animals, later in life, display a typically male behavior pattern." Could something common in popular brands of baby bottles be modifying sex differences? "Exposure to very low doses of bisphenol-A results in masculinization of the female brain," the article went on, quoting one of the study's co-authors, Ana M. Soto. Contrary to popular expectations of male-female differences, female mice are usually the more avid explorers, whereas male mice are more passive when confronted by novel situations. Perinatal bisphenol A exposure, even at extremely low doses, made adult females more timid and unwilling to explore, a typically masculine behavior for mice. In addition to masculinizing female mouse behavior, the chemical also changed female mouse brain structure, reducing the size of a part of the brain that is usually larger in females.[1]

The media attention on bisphenol A's effects on sex differences in rodents suggests some of the complexities of gender in modern America. Endocrine disruptors can indeed have profound effects on the processes of sexual development, and anxieties about sex and gender have stimulated much of the popular concern about the synthetic chemicals. But because the expression of sex differences is intertwined with cultural norms, identifying the risks posed by endocrine disruptors becomes complicated. As a thought experiment, imagine that the media had proclaimed that common pollutants undermine natural differences between races, making Caucasian men act more like African American men and vice versa. Most people would immediately question the definition of "natural" differences between races. Most would also understand that cultural factors are key elements shaping the ways differences are expressed at any particular time in any particular culture. Yet these cultural considerations would not mean that the pollutants themselves were harmless. Rather, they would mean that understanding their risks would involve careful consideration of the boundaries between nature and culture. Both skin color and gonad structure have a biological basis. For both race and sex, however, culture shapes the ways that particular societies group biological characteristics into separate categories.

In 1993 the biologists Richard Sharpe and Niels Skakkebaek published a paper in the *Lancet,* the premier British medical journal, that opened with an attention-gripping abstract: "The incidence of disorders of development of the male reproductive tract has more than doubled in the past 30–50 years while sperm counts have declined by about half. Similar abnormalities occur in the sons of women exposed to diethylstilbestrol (DES) during pregnancy and can be induced in animals by brief exposure to exogenous oestrogen/DES during pregnancy. We argue that the increasing incidence of reproductive abnormalities in the human male may be related to increased oestrogen exposure in utero, and identify mechanisms by which this exposure could occur."[2]

Although exposure to estrogenic endocrine disruptors was only one of five possible mechanisms suggested by Sharpe and Skakkebaek, it captured the attention of researchers, funding agencies, and the popular media. In the 1980s, when the zoologist Theo Colborn had first suggested that synthetic chemicals might mimic hormones and contribute to reproductive problems in wildlife, she had a difficult time interesting funding agencies in her research. But after the biologist Lou Guillette went to Congress and announced, "Every man sitting in this room today is half the man his grandfather was, and the question is, are our children going to be half the men we are?" funding for such studies materialized. Popular books such as *The Feminization of Nature* and scientific articles with titles such as "Masculinity at Risk" and "The Human Testis—An Organ at Risk?" suggest potent cultural anxiety over norms of sexual difference.[3]

Reacting to such headlines, some cultural theorists dismissed concern with endocrine disruptors as reflecting little more than essentialist assumptions about sexual difference and development. But we cannot easily dismiss endocrine disruptors as a manifestation of cultural anxiety and therefore not of concern to material bodies. Reproductive cancers linked to pollutants are not mere anxieties; they have material reality. Intersexual and transsexual conditions correlated with chemical exposures are not simply what the theorist Judith Butler calls performances playing with cultural norms of gender. As the gender theorist Celia Roberts argues, while we need to question assumptions about the essential nature of men and women that underpin some interpretations of endocrine disruptors, we also need to recognize that "there are strong biological effects on our bodies and activities." We share kinship with other animals, which "does not mean that we are 'fixed' or biologically determined in any simple way," but biology still shapes us, as does culture.[4]

Bodies are, in effect, complex ecosystems made up of a dynamic interweaving of material and cultural feedbacks that are themselves subjects and sources of environmental degradation. Whatever humans do to the natural world finds its way back inside our bodies, with complex and poorly understood consequences. And in turn, what happens inside our bodies makes it way back into the broader world, often with surprising effects.

Sexual development appears to be particularly vulnerable to chemical disruption, but sex is not a simple matter for humans, or for other vertebrates. Although *sex* typically refers to biological sex, whereas *gender* refers to cultural constructions (a person's self-representation as male or female), sex and gender are not entirely separate. As a report by the Institute of Medicine on the biology of sex differences puts it, the two are "part of a single system in which social elements act with biological elements to produce the body." Exogenous factors—things outside the body—constantly find their way into endogenous pathways, shaping differences between male and female at the genetic, phenotypic, and cultural levels. Social interactions, for example, can change hormone levels, which can in turn affect bone deposition and mass.[5]

For thousands of years, people have been puzzling over how sexual differences develop in the womb. Our current understandings of sex differences are historical products of a long interweaving of science, cultural understandings, technology, religious beliefs, and the materiality of biological systems. As our social views have changed, so too have our scientific views of the development of sex and gender differences. All this complicates any approach to endocrine disruption because our ideas about "normal" and "natural" differences between the sexes color the way we look at hormonal effects.

The historian Nelly Oudshoorn notes that until the eighteenth century, most Europeans thought of male and female bodies as essentially similar. Women were believed to have the same genitals as men, hidden within their bodies. In the nineteenth century, with the development of medical and scientific technologies that revealed more of the body to the scientific gaze, male and female bodies were seen as opposites. Until the discovery of chromosomes and hormones, the womb and the ovaries were believed to be the seat of femininity.[6]

Fascinated by chromosomes, researchers debated whether sex was predetermined. Some researchers believed that the embryo had an indeterminate sex, and it was the environment rather than the chromosomes that determined whether it would develop as male or female. During the late nineteenth century, for example, many scientists believed that nutrition during pregnancy determined sex. If a mother ate well during her first trimester, a boy would be the result, whereas poor nutrition would make the fetus develop into a girl.[7]

New technologies for exploring genes dramatically changed understandings of sexual differentiation in the latter half of the twentieth century. The search for a "testis-determining" gene on the Y chromosome became something of a holy grail for many developmental biologists. In the 1950s the researcher Alfred Jost argued that embryos develop in a non-sex-specific manner until, in XY fetuses, something causes the "indifferent gonad" (which has the potential to develop into either male or female gonads) to become a testis. The testis then secretes hormones which masculinize the reproductive tract, creating masculinity. In the 1980s, a gene named SRY (for Sex-determining Region Y) was isolated from the human Y chromosome and declared to be the gene that confers maleness on the developing fetus. The SRY gene dominated sex-determination research for a decade after its discovery. The sociologist of science Joan Fujimura writes, "As many feminist writers have pointed out, the development of females appears to be discussed by biological and medical texts in terms of passivity—in the absence of an active trigger required to induce male development, an embryo develops ovaries, a female secondary sexual characteristic." Many geneticists conceptualized femaleness as the "default pathway," a passive absence of maleness, whereas they saw maleness as an active process determined by a single powerful gene.[8]

In the twenty-first century a view of sex determination and sexual differentiation emerged that included genomic, hormonal, and environmental influences. Researchers currently argue that sexual difference begins in the genome with the X and Y chromosomes, known as the sex chromosomes, but it does not end there. The sex chromosomes make up only about 5 percent of the total human genome, which means that the vast majority of male and female genes are the same. Variability within the same sex in most traits is far greater than variability across sexes.[9]

While a common assumption is that chromosomes determine whether a person will be male or female, sex determination is more complex. The genotype (the inherited instructions carried within the genetic code) does not directly determine the phenotype (the observable characteristics of an organism, such as testicles or breasts). A person's genotype is made up of many genes, most of which remain inactive. Because only a tiny subset of a person's DNA sequence is expressed, the regulation of gene expression is essential to every stage of development. (Gene expression is the process by which inheritable information from a gene is made into what is called a functional gene product, such as protein.)

Which cells turn into which organs and how those organs function depend on a suite of interdependent systems. Hormones influence gene expression, and hormones in turn are influenced by genetic, environmental, and epigenetic factors. It is not so much the genes themselves that differ between males and females; it is the ways those genes are expressed. Questions of gene expression blur the traditional boundaries between gene and environment, along with the distinction between self and nonself, because patterns and processes of gene expression immediately draw the environment into the processes that direct prenatal development.[10]

In mammals, the presence or absence of the Y chromosome is only one of several factors that initiate the direction of sexual differentiation. Genes are part of the complex systems that guide sexual differentiation, but the process is more complicated than one in which a single male SRY gene determines maleness, as many researchers once believed. At least seventy different genes, numerous hormones and proteins, and receptors for those hormones regulate sexual differentiation. Cells communicate their messages by secreting proteins, and these signals are vulnerable to chemical disruption, making the period when differentiation is taking place a time when the embryo is particularly susceptible to damage from toxic exposures.[11]

Early in development, the indifferent gonad can differentiate into either ovaries or testicles, depending on hormonal signals. If a Y chromosome is present in the embryo, the indifferent gonad differentiates into testes, which then produce the steroid hormones that lead to the formation of male reproductive structures. By the sixteenth week of gestation, the XY fetus's testosterone levels have become quite high—comparable to that of adult men. These weeks of high testosterone levels are critical for masculinization of the developing fetus, and hence vulnerable to disruption from synthetic chemicals such as DES. For example, DES exposure may affect the Sertoli cells in male fetuses, which control the development and descent of the testes and the development of sperm. Problems with Sertoli cells can lead to cryptorchidism, in which the testicles fail to descend.[12]

Development of external genitalia in the male fetus is initiated by a process called the androgen-signaling cascade, which involves testosterone regulation and synthesis. An enzyme converts testosterone in the skin and genital tissues into a second hormone called dihydrotestosterone, which signals the development of external male genitalia. Between the eighth and twelfth week of gestation, the genital tubule develops into the penis, while the urethral tube fuses in the midline, becoming the scrotum. Endocrine disruptors can interrupt these steps. DES exposure, for example, may alter the fusing of the urethral tube and lead to hypospadias, in which the urethra opens not at the tip of the penis, but along the scrotal region instead.[13]

The process of sexual development in female fetuses also involves a complicated set of chemical messages that signal the indifferent gonad to develop into reproductive structures. In the 1990s researchers presented evidence that a separate gene, DAX-1, was involved in female sex determination. The subsequent discovery of numerous additional genes with cascades of genetic switches and expressions quickly complicated the view of female sexual development.[14] Differentiation of the external female genital tract begins at about the eighth week in female fetuses, and the ovaries begin to develop at about the eleventh week of gestation. By week twelve, the genital tubule develops into a clitoris, while the urogenital membrane develops into the labia. The formation of the oviduct, the uterus, the cervix, and the upper part of the vagina in XX fetuses is partly guided by genetic information, but these developmental stages depend on changing hormonal signals that tell various genes to turn on and off. Chemical exposures can disrupt these signals, leading to a variety of reproductive problems later in life. For example, in female laboratory animals, prenatal exposure to DES can suppress a gene named MSX2, resulting in vaginal, uterine, and fallopian-tube problems similar to problems experienced by women who were exposed to DES in the womb.[15]

Understanding the potential for synthetic chemicals to cause birth defects requires more than medical research alone, for defining "normal" and "abnormal" is a social decision, not just a medical decision. When developmental changes in sexual characteristics have been linked to pollutants, distinguishing normal from abnormal is even more problematic. For scientists, DES has become a model for the long-term effects of fetal exposure to an endocrine disruptor. Yet it is not entirely clear what this model is telling us. Some researchers interpret the DES story as evidence that endocrine disruptors are a grave threat to sexual differentiation. Others interpret it as evidence of the opposite, arguing that DES experiences suggest that endocrine disruptors pose little risk.[16]

The assumption in both perspectives is that stable boundaries exist between male and female, synthetic and natural, and endogenous and exogenous hormones. Consider the cyclist Floyd Landis, who was accused of illegal steroid use after his record-breaking win in the 2006 Tour de France. When Landis failed two drug tests, officials (and most of the public) decided that he must have taken steroids. The first test found an unnaturally high ratio of testosterone to epitestosterone, and the second test found evidence of synthetic testosterone in his blood. Landis eventually lost his title, even though he continued to insist that the test results were open to question. The first test assumes that a fairly stable balance exists between two hormones, and the second test assumes that we can easily tell the difference between natural and synthetic steroids. Hormone levels and ratios, however, are not stable in our bodies; social interactions can quickly change them. Testosterone levels may rise in a male bird immediately after a competition with another male, for example, and a similar pattern exists in human athletes.[17]

More surprising, perhaps, the boundary between natural and certain synthetic hormones is not clear. Testing for synthetic steroids relies on the ratio of carbon 12 to carbon 13 isotopes, but the baseline for normal used in the French tests assumes a European diet, which tends to be relatively low in carbon 12. African and some American diets are richer in foods such as soy, yam, and millet, which contain natural steroid hormones that have a relatively high level of carbon 12. Moreover, synthetic steroids are typically derived from soy, a source of plant estrogens that laboratories convert to synthetic testosterone. The problem is that not only laboratories do this conversion; people who eat high quantities of soy can alter their body's production and breakdown of estrogen, testosterone, and other steroids. Athletes who eat substantial amounts of soy and yams could theoretically appear to have synthetic testosterone in their blood.[18] The key point here is not that Landis might be innocent but that our assumptions about natural and synthetic, exogenous and endogenous, and masculine and feminine begin to fall apart when we consider steroid hormones.

No simple correspondence traces a clear line from chromosomes to sex to gender. Hormones influencing sexual development come partly from the mother, partly from the fetus, and partly from the external, chemically saturated world, and all mediate these complex transformations. Steroid hormones are not just male or female hormones: they transform from one to the other, and both sexes produce and respond to both.

One focus of research on endocrine disruptors has been their effects on sexually dimorphic behavior. In this work researchers have tended to assume that sexual dimorphisms, including brain, and hence behavior, result from hormones produced by the developing gonad. According to this model, exogenous hormones such as endocrine disruptors could disrupt sex differentiation by mimicking the action of endogenous hormones on the developing gonads and the developing brain. And indeed, some of these effects have been found in laboratory animals, in which males exposed to DES tend to display behavior more typical of the females, and vice versa.[19]

Some researchers have argued that if endocrine disruptors were really a problem in sex differentiation, sons exposed to the feminizing influence of

DES should display more typically feminine behavior, such as playing with dolls, whereas daughters exposed to synthetic androgens should display more typically masculine behavior, such as doing well at math. Because researchers have failed to find these predicted relations between DES exposure and gendered behavior, some review articles conclude that endocrine disruptors are not a problem for human development. This logic assumes that fetal exposure to sex hormones determines gendered behavior in children and adults, yet feminist critics and developmental biologists have shown that this model is oversimplified.[20]

Debates about what determines gender differences in behavior are not new. Frank Beach, the leading scholar in sexual behavior from the 1930s through the 1950s, argued that both biology and social learning influenced behavioral differences between males and females. Although he saw biology as important, he did not consider it preeminent. He argued that various physiological systems, including the hormone and the nervous system, played a part in shaping behavior and that individual differences were critical for understanding variation between sexes. Beach's research also explored the effects of experience, context, and social history on sexual behavior.[21]

Beach's views of the complexity of sexual difference were challenged by the anatomist William C. Young, who published work in 1959 showing that hormones given to fetal guinea pigs influenced sexually dimorphic behavior throughout life. The research stimulated by this paper has often focused on the central role of biology, rather than culture or the relation between the two, in shaping sex differences. As the historian and biologist Anne Fausto-Sterling argues, Young's influence in the field led to decades of hormone research that tended to essentialize differences between males and females, implying that hormones determined behavior and that genes determined hormones.[22]

Feminists and scientists alike have rejected such arguments. But as Roberts points out, "The danger to feminist criticisms of scientific positions such as those describing the role of hormones in producing sex differences, is that if biology is simply rejected as essentialist, simplistic, or just plain unbearable, it is reinstated as unknowable and beyond the social, beyond feminism." Seeing hormones as essentialist reaffirms the social/biological, natural/cultural dichotomies. Fausto-Sterling urges us to "find a way to talk about the body without ceding it to those who would fix it as a naturally determined object existing outside of politics, culture, and social change."[23]

Important as it is to pay attention to the social constructions of sexual difference when considering the role of hormones, we must not ignore empirical observations, such as the startling discovery that more than 30 percent of fish surveyed in many British streams are now intersex. The rates of occurrences of intersex conditions, along with those of many other reproductive problems, from infertility to hypospadias to cryptorchidism to double uteruses to blind vaginas to uterine tumors to endometriosis, all appear to be increasing. Much of that increase is probably due to new detection methods such as ultrasound, karyotyping, and genetic techniques that allow researchers to detect variations that may have been present but gone unrecognized in earlier decades. At least some of the increase, however, cannot be easily explained by changes in detection or reporting.[24]

Intersex conditions are certainly part of the natural range of variability.[25] Research that suggests industrial and pharmaceutical pollutants might be increasing the rate of intersex conditions has been controversial within some intersex activist communities, many of whom are concerned that it will support fundamentalists who see such conditions as defective, immoral, and in need of surgical correction. Other activists argue the opposite, believing that research on chemical links to intersex conditions has the potential to increase public awareness and acceptance of those conditions. The Intersex Society of North America, the largest intersex activist group, has begun working with birth defects research groups to investigate possible links between toxins and intersex conditions. In part, the members want to have a say in defining the research agenda, rather than simply being objects of research. And in part they want to shift the debate from morality to biology. But the fear remains that environmental health research will benefit only the scientists: that endocrine-disruption results will be used to characterize intersex individuals as freaks of nature who need to be "cured."[26]

Sexual difference, however, is never simple in humans, or in any other vertebrate species. Sexual difference is not constructed by genes or the steroid hormones in the fetal environment or the mother's immune system or synthetic chemicals in the fetal or maternal environment or the postnatal culture alone. All these elements are dynamic, and all interrelate, acting within themselves and upon each other. As Fausto-Sterling argues, sexual difference is part of a "*gene-environment system* in which genes only assume importance when they respond to a particular environment and a particular environment changes the body by activating sets of genes."[27]

Vertebrates have a remarkable range of individual variation in sexual characteristics. Among animals other than humans, nearly every vertebrate species studied responds to endocrine disruptors by increasing the frequency of intersex conditions, both in the laboratory and in wildlife populations, and we should not be surprised to find something similar occurring in humans. Yet it is important to remember that although synthetic endocrine disruptors may affect this variation, they do not create the variation. Intersex variations are not defects caused by synthetic chemicals; they are part of the exuberant diversity of vertebrate evolution.[28]

A species' pattern of sexually differentiated behavior is influenced by environment, by historical accident, by evolutionary history, by social interactions, and by hormones. One of the insights of evolutionary ecology is that environmental and social conditions influence the evolution of sexual differentiation, often very rapidly. Evolutionary theory is not about genes controlling behavior; it is about changing strategies that make sense in particular ecological and social contexts. This is helpful for thinking about sex differences, because instead of normalizing differences, evolutionary theory allows us to envision sexual differences as historical constructs, shaped by genes, behavior, ecology, and environment.[29]

If a trait has a genetic or hormonal component, that does not mean that it is therefore fixed and unchanging. Many laypeople tend to assume that genes and hormones are the blueprints that determine behavior. Yet often influences go the other way, with behavior shaping genes, rather than genes shaping behavior. In the 1980s, researchers in the laboratory of the endocrinologist John Wingfield showed that behavioral interactions influenced steroid hormones, which then influenced brain structure. In white-crowned sparrows, song centers — the regions in the brain that correlate with song types — differ between male and female birds. Males, which sing courtship songs, have different brain structures from females, which do not. The song that a bird sings influences its hormones, and those hormone changes then alter the bird's brain structure. Feedback mechanisms are much more complicated than had been assumed in the simple model of genes determining hormones, which then determined sexually dimorphic behavior. Researchers in Wingfield's laboratory learned that certain male birds had more testosterone and were also more aggressive. The media tended to report this finding as proof that testosterone causes male aggression. But in many birds it was the behavior that influenced the level of testosterone circulating in the bloodstream, rather than vice versa.[30]

Even bone differences between males and females could be affected by behavior. Using supplemental-feeding experiments, I worked with a group of ecologists to create environmental fluctuations that differentially affected male and female blackbirds. Early in the breeding season, females received supplemental food on their wetland breeding territories, while males did not. These seemingly minor food differences led to rapid changes in reproductive behavior (and presumably hormones), which then influenced the degree of sexual dimorphism between male and female bone lengths. These changes occurred quickly: over the course of a six-year study, we found significant changes developing in bone sizes and sexual dimorphism.[31] Birds as well as people support Fausto-Sterling's argument that social interactions and culture shape even the hard substance of bones.[32]

Diethylstilbestrol, like both natural and synthetic steroids, changes the rates of these evolved variations in sexual characteristics, but not in simple or direct ways. Although it demasculinizes certain aspects of male development, DES appears to do the opposite in female development. This might seem paradoxical: if estrogens feminize fetuses, would not a synthetic estrogen feminize a girl? But steroids are not simply male hormones or female hormones, and they are not fixed or stable in our bodies. Estrogen can be transformed in our bodies to testosterone, which can then be changed back to estradiol. Typically, a developing female fetus is protected from the influences of circulating maternal estrogens by a protein called alpha-fetoprotein, which prevents them from reaching the fetal brain. A male fetus, however, produces testosterone, which converts to estrogen, and it is this estrogen (not the testosterone) that masculinizes the developing male fetal brain. A female fetus is not typically exposed to the masculinizing effects of estrogen because alpha-fetoprotein has bound up the mother's estrogen, and the fetus is not producing testosterone that could be converted to estrogen. But endocrine disruptors such as DES and bisphenol A have a much lower affinity for plasma-binding proteins. Thus the brain is not protected from DES and bisphenol A in the same way that it is protected from the mother's circulating estrogens, enabling the endocrine-disrupting chemicals to have effects on the differentiating brain similar to those of androgens.[33]

As this example suggests, endocrine disruptors often do not act directly on the developing fetus; instead, they act indirectly, changing the networks that structure the development system. They might change hormone binding, metabolism, or conversion from estrogen to testosterone, or they could alter epigenesis and methylation of various genes that structure hormone synthesis. All these can indirectly change the actions of endogenous hormones. The chemical industry has taken advantage of this complexity, using it to argue against regulation. After a research team headed by Scott Belcher discovered in 2005 that very low doses of bisphenol A change estrogen's rapid-signaling mechanisms and alter fetal brain tissue, a representative from the American Plastics Council pointed out that it was not possible to say whether this was a good or a bad thing for fetal brains: "It's very complex systems that they're looking at." In the view of the chemical industry, such complexity makes it impossible to regulate the chemicals.[34]

When we point to the complicated and fluid boundaries between male and female, we run the risk of implying that environmental contamination is merely a social construct. Nothing could be farther from the truth. The endless series of reproductive problems among the women in my family were hardly social constructs. The scalpels that sliced out my uterus existed within a network of social signifiers, but the blood that flowed had material reality. As the environmental scientist Joe Thornton points out, if drinking a poison kills you five minutes later, that's an acute effect, and it is fairly easy to understand because the compound quickly produces a dramatic change in the state of the organism. But chronic, low-level effects are much more difficult to identity or measure, much less regulate. Thornton writes: "There is no single 'normal' state for any of these functions, all of which vary naturally within some range in the population. This natural variability greatly increases the noisiness of the results and reduces a study's power to establish a statistically significant association of exposure with effect."[35] When there is no single normal state, how do we decide when something has changed, and how can we identify a change as problematic?

A new ecology of health can help us unravel these complexities. In an ecological approach to health, we recognize that the body is enmeshed in a web of relationships, not isolated within a castle whose threshold can only be breached by a sustained attack from the outside. In *Having Faith*, the ecologist and writer Sandra Steingraber calls for us to stop imagining bodies as separate from environments. Steingraber uses her own pregnancy as a way to explore the biological, ecological, and political interconnections of reproduction. The uterus, she argues, is an ecological habitat akin to an inland ocean. As the sociologists Steve Kroll-Smith and Worth Lancaster write, "Surrounding this ocean is the body of the mother mediating between womb and an external nature increasingly saturated with industrial poisons."[36] A healthy pregnancy does not exist in static, disconnected space; rather, it is a complex set of systems, forged by connections between the mother and developing fetus, between genes and hormones, between past generations and future generations.

In 2007 an article appeared with the title "Epigenetic Programming by Maternal Behavior and Pharmacological Intervention: Nature Versus Nurture; Let's Call the Whole Thing Off."[37] The subtitle expresses one of the key insights we can gain from epigenetics, which may finally make us move away from dichotomies between nature and culture. Our bodies exist in dynamic ecological relationships with the environments in which we live. It is easy to fall into the trap of thinking that we are isolated individuals constantly bombarded by synthetic external disturbances we need to fight off. But physiologists have moved away from seeing organisms as separate individuals; they now sound more like the postmodern theorist Donna Haraway when they write about an organism as "an interaction between a complex, self-regulating physiological system and the substances and conditions which we usually think of as the environment." Because of its ability to alter reproductive systems, hormone systems, and even immune systems, DES undermines our belief in an individuated body, or what Haraway has called "an individual defended by the immune system against non-selves." Haraway begins with the immune system because it has classically been conceived of as the ultimate defender of self against nonself. Haraway suggests that we might better understand ourselves as a "semipermeable self able to engage with others (human and nonhuman, inner and outer), but always with finite consequences; of situated possibilities and impossibilities of individuation and identification; and of partial fusions and dangers."[38]

None of us is an isolated individual. We are all networks of self and nonself, of our own identities and DNA interwoven with the colonies of parasites and bacteria and viruses that make up our bodies. Our deepest sense of self reflects our personal, cultural, and evolutionary histories, including the viruses that millions of years ago were, in immunological terms, "nonself invaders" but which eventually became incorporated into our DNA and now modulate our responses to the hormonal webs within which we exist. In "A Cyborg Manifesto," Haraway notes that "a stressed system goes awry; its communication processes break down; it fails to recognize the difference between self and other. . . . Among the many transformations of reproductive situations is the medical one, where women's bodies have boundaries newly permeable to both 'visualization' and 'intervention.'"[39] We live in this world now, where our hormone system, one of the two key communication systems in our bodies, is certainly going awry, complicating the divide between self and other, and creating permeable boundaries that are increasingly vulnerable to chemical contamination.

Haraway's ideas challenge what Kroll-Smith and Lancaster call the "Enlightenment-inspired idea that bodies and environments are genuinely discrete realities." They write: "It is customary to think of bodies and environments as if they are two separate and distinct entities. The pronoun 'my' in front of 'body' signals a possessive interest in human bodies that differs from the typical article 'the' that precedes 'environment.' 'My body' and 'the environment' are lingual signals that two ontologically discrete things are being discussed. In a world enunciated with discrete categories, all of us simply know where our bodies end and the environment begins." Risk assessments, they argue, rest on these ontological assumptions of separateness: "By assuming a categorical distinction" between bodies and environments, regulatory authorities can then "issue a 'pollutant discharge permit' licensing the right to contaminate environments as 'long as the exposure is below the *threshold* at which' environmental toxins adversely affect bodies."[40] The implicit assumption is that bodies and environments are separate enough that a toxic chemical can contaminate the soil, water, or air without contaminating people.

In contrast, ecological models of health require that we envision the body as permeable to the environment, just as earlier generations of physicians did. Microbial ecologists have long seen the body as an ecosystem. The microbiologist C. Tancrède argues that "the human host and its microbial flora constitute a complex ecosystem whose equilibrium serves as a remarkable example of reciprocal adaptation."[41] The ecologist Craig Allen's work explores similarities between immune systems and ecological systems, focusing on how "resilience is maintained by complex systems under the threat of invasion." Like most ecosystems, the body undergoes disturbances from natural toxins, parasites, solar radiation, and mutagens. Health is not the absence of stressors; it is the ability to respond to these stressors. The immune response and mechanisms of cellular and DNA repair are all part of a complicated ecosystem that regulates and repairs the human body. Multiple interactions, feedback loops, and nonlinear system dynamics shape the health of individuals, populations, and ecological communities. Reductionist science and linear causality are useful approaches for many questions, but they falter when understanding the dynamics of complex systems.[42]

René Dubos argued decades ago that health can be viewed ecologically not as the simple absence of disease but rather as the ability to adapt to changing circumstances. Health, Dubos believed, consists of adaptations to stress, feedback loops, pathways of nutrients and energy, flows of energy and waste, and regulatory mechanisms that control these relations. As Thornton writes, "Compromised health may become apparent only when new sources of stresses are applied and the individual fails to adapt."[43]

Traditional medicine and public health practices have been reductive, focused on individual risk factors for disease. This approach has been successful for understanding major causes of certain ailments such as lung cancer and heart disease. Other physical problems, however, such as birth defects, premature births, miscarriages, and many reproductive cancers, are resistant to an approach based on individual risk factors. The physician and public health researcher Ted Schettler suggests instead that we view diseases as "ecological manifestations of multiple changes in the dynamic system in which people are conceived, develop, live, and grow old." Thornton writes, "Bodies are resilient things. There are poisons in nature, but organisms have ways of coping with them. Their bodies repair injuries, fight off infections, degrade and excrete foreign compounds, and maintain a steady internal state despite fluctuations in the environment." Endocrine disruptors "can disrupt the very processes that make organisms resilient in the first place." They can "compromise the body's ability to defend itself against infectious disease and interfere with the hormones that coordinate" reproduction.[44] To address complex health problems, we need to better understand their ecology, not simply try to reduce individual risk factors.

Endocrine disruptors such as DES and bisphenol A should concern us not simply because they have the potential to harm the body, for bodies are constantly adapting to substances that can cause harm. The disturbing thing is that endocrine disruptors transform the body's ecological repair mechanisms, often at the biochemical level. In particular, they alter the epigenetic processes that link environment and gene, leading to changes in gene expression and in turn to changes in the numbers and types of immune cells in the blood, as well as changes in hormone production and metabolism. They alter ecological processes of human health, just as they alter broader ecosystem processes. They transform the network of responses that animals have evolved in their adaptation to stressors. Amphibians, for example are often exposed to parasites such as trematodes, and throughout their evolutionary history they have evolved a set of responses to these parasites. Exposure to endocrine-disrupting pesticides such as the common herbicide atrazine changes their ability to respond. In 2002, the ecologist Joseph M. Kiesecker linked increased trematode infection and limb deformities to pesticide exposure.[45] Trematodes alone may not harm a developing frog. Pesticides alone may not harm the frog much, either. But in the presence of trematodes, the pesticides can disrupt the frog's ability to respond to parasites and other threats in its ecosystem, causing limb deformities. Similarly, research by David Skelly's laboratory at Yale shows that intersex frogs are more common in suburban areas than in forested, undeveloped areas. Agricultural areas have intermediate levels of intersex frogs, suggesting that common patterns of land-use change, where suburbs are replacing both fields and forests, may be harming amphibian reproductive health.[46]

Chemical pollutants change the network of genetic, immunological, neurological, hormonal, and environmental interrelationships that control sex and reproduction in vertebrates, and this can kill an individual, eliminate a population, or drive a species extinct. Understanding endocrine disruptors means reconsidering our bodies and our identities, seeing them not as separate isolated objects but rather as what Bruno Latour termed hybrid networks. The material and the cultural, in Latour's terms, "weave our world together," yet these weavings have often become invisible to us.[47]

Anne Fausto-Sterling writes that we have "forced the hybrid networks linking nature and culture underground. . . . Although a strategy of ignoring hybrids worked in the beginning, it embodied a paradox. The better it worked, the more unacknowledged hybrids developed. The more we dominated nature, the more the proof of our domination poured into culture; the more culture dominated nature, and the more we created objects that were neither truly natural nor truly cultural."[48] The hybrids we have created with endocrine disruptors resist our attempt to define clear boundaries between natural and synthetic, and between male and female. Try as we might to relegate these hybrids to the world of monsters, freaks, five-legged salamanders, two-headed monkeys, and sex-changing salmon, we cannot, for they are increasingly ourselves.

CHAPTER 9

Precaution and the Lessons of History

Questions about risk, profit, and the burden of proof have troubled U.S. regulatory agencies ever since Harvey Washington Wiley called for a version of the precautionary principle in the early decades of the twentieth century. Since 1998 they have coalesced around the demand for a precautionary approach that would place the burden of proof on those who profit from toxic chemicals. That year, thirty-two scientists and physicians concerned about endocrine disruption published a consensus statement known as the Wingspread Statement on the Precautionary Principle. They wrote: "When an activity raises threats to the environment or human health, precautionary measures should be taken, even if some cause-and-effect relationships are not fully established scientifically. In this context, the proponent of an activity, rather than the public, should bear the burden of proof."[1] Yet the precautionary principle is not easy to implement, for the environmental or health risks of a particular action are usually uncertain and occur in the future, while the costs of averting it are often immediate.

Precaution has a long history in public health. As Sonja Boehmer-Christiansen argues, a formal precautionary principle evolved out of the German concept of *Vorsorgeprinzip,* which developed in the legal tradition of the 1930s democratic socialism. Vorsorgeprinzip centered on the concept of good household management, a concept that justified state involvement in planning economic, technological, moral, and social initiatives.[2]

Precaution had been adopted well before this in public health efforts, however. When the British physician John Snow recommended removing the handle from the Broad Street water pump in an attempt to stop London's 1854 cholera epidemic, that was a form of precaution. Scientists were still uncertain of the causes of cholera when Snow acted. He had found a correlation between polluted water and cholera five years earlier, but most scientists and physicians rejected his thesis as untenable, believing that airborne contaminants caused cholera. The biological mechanism underlying the link between polluted water and cholera was unknown. Yet even without firm proof, Snow had enough information to judge that the possible costs of inaction would probably be greater than the costs of action.[3]

Snow's vision of protecting the public through precautionary action continued as an important thread in public health. During the first decades of the twentieth century, Harvey Wiley and Walter Campbell argued that the federal government needed to use precaution as the basis of regulation. The Food and Drug Administration, they believed, needed to sift evidence from multiple perspectives, not just the industry standpoint, to find preliminary evidence that might suggest possible links between a compound and an adverse, potentially irreversible, outcome. This preliminary evidence might come from experiments on animals or from structural similarities between a given chemical with unknown effects and one with known effects. Precaution was justified, they believed, when the potential costs were high or irreversible compared to the benefits, when the person who bore the costs did not receive the benefits, and when preliminary evidence suggested a possible link between an action and a harm, even when the exact biological or chemical mechanisms underlying that link were still uncertain.[4]

Beginning in the 1970s, scientists and activists made efforts to extend the idea of precaution from the public health arena into broader environmental decision-making. In the 1970s, German foresters struggled to establish the causes of dying forests and developed a precautionary principle similar to that proposed by Wiley and Campbell. The 1985 report on the German Clean Air Act noted that a precautionary approach requires more than establishing the "level of proof needed to justify action to reduce hazards (the 'trigger' for action)." Other important elements include monitoring for early detection of hazards; promoting alternatives such as clean production and innovation in green chemistry and engineering; "the proportionality principle, where the costs of actions to prevent hazards should not be disproportionate to the likely benefits"; and a commitment to take action "before full 'proof' of harm is available if impacts could be serious or irreversible." The 1992 Rio Declaration on Environment and Development was explicitly grounded in precaution, and the principle has since become central to consumer and environmental protection policy in the European Union. In 2007 the European Union passed a law mandating that chemical companies must demonstrate that their products are safe before they can be placed on the market.[5]

Industry has generally opposed efforts to extend precaution from medical to environmental policy, fearful that such an approach could stall innovation. The Business Roundtable was founded in 1972 to represent two hundred of the nation's largest corporations, and this association has taken an increasingly active role in opposing environmental regulation. Gerald Markowitz and David Rosner argue that the association's strategy has been to accentuate elements of complexity and uncertainty and then to argue that "economic interests should not be challenged until science has proven danger. Precaution is equated with economic and social stagnancy. . . . Progress, as defined by the industrial community, trumps precaution."[6] By the mid-1970s, as consumer concern over environmental pollution placed increasing pressure on industry, it responded with a "frontal assault on the public health ideals of prevention," hiring product-defense firms, public relations agencies, and scientists who "systematically attacked environmentalists and labor activists as luddites determined to stifle our economy."[7] Industry advocates sought to portray precaution as a novel and reckless idea, rather than a long-held principle at the heart of public health. What was most daring about this campaign was industry's largely successful effort to rewrite history in the public eye, portraying precaution as a new idea and indisputable proof of harm as a historical precedent.

Histories are not just academic exercises; they are political acts. Historians tend to be reluctant to work with policy makers for fear that they might be accused of the historian's cardinal sin: presentism, or the error of judging past actions by the standards of the present. But we can and should learn from the experiences of the past, and we can do so without falling into presentism. Policy making is often based on arguments about the past, although rarely are those arguments as explicit as industry's attempts to rewrite the history of the precautionary principle. Foresters, for example, observe the ways forests have responded to particular ecological disturbances or silvicultural treatments in the past. They use their observations to derive hypotheses about how today's forests might respond to various changes, including logging, climate change, and suburban development. Hospital review boards examine past medical mistakes to avoid repeating them. But environmental policy makers rarely take a formal or structured approach to examining historical case studies to learn when evidence emerged about potential risks or when implementation of policies might have saved lives or suppressed innovation. As the political scientist Richard Neustadt and the historian Ernest May argue, policy makers need to be more explicit about framing hypotheses and testing them with historical evidence, rather than relying on anecdotes about the past.[8]

In 2001, a European Union team charged with implementing the precautionary principle examined fourteen case studies of historical hazards. The case studies involved an agent (such as mercury) that most contemporaries had regarded as harmless at prevailing levels of exposure until additional evidence about harmful effects emerged. The goal of the exercise was to identify when the first credible "early warnings" of potential harm emerged, determine how regulatory authorities responded (or failed to respond) to those warnings, and calculate the resulting costs and benefits of that inaction. The team came up with several lessons for policy makers that correspond closely with lessons we should learn from the case of diethylstilbestrol.[9]

One critical lesson discussed in the European Union case studies concerns the importance of first recognizing limits to knowledge and then accepting that continued uncertainty is no justification for inaction. As the European Union team writes, "No matter how sophisticated knowledge is, it will always be subject to some degree of ignorance. To be alert to—and humble about—the potential gaps in those bodies of knowledge that are included in our decision-making is fundamental. Surprise is inevitable." In *Seeing Like a State,* the political scientist and anthropologist James C. Scott describes the ideal land manager as one who is humble, experienced in making mistakes, willing to be wrong, and pragmatic about the limits of his or her knowledge. The same can be said of an ideal regulator, who should be willing to acknowledge the possibility of surprise. The European Union team notes that "acknowledging the inevitable limits of knowledge leads to greater humility about the status of the available science, requiring greater care and deliberation in making the ensuing decisions. It also leads to a broadening of appraisals to include more scientific disciplines, more types of information and knowledge, and more constituencies."[10] The regulators involved with DES understood that their knowledge about the actions of synthetic hormones was limited, but when it came time to assess risks and make decisions, they seemed to lack the humility that a partial understanding requires.

Regulators need to look beyond direct impacts from a chemical and anticipate "as wide a range of conditions and effects as can reasonably be anticipated. Whilst accepting that even the broadest appraisal processes may still fail to foresee 'surprises,' there is much that can be done to guard against some of the consequences of the ubiquitous experience of ignorance and surprise."[11] Civil engineers learned this lesson a long time ago; it is now accepted engineering practice to assume a degree of ignorance and devise strategies to prevent outcomes that are by definition still unknown.

Ignorance can sometimes be intentional. An industry might prefer not to find out about the potential harm its product might cause because continued uncertainty means continued profits. Without monitoring of potential hazards, we are almost guaranteed to be more ignorant than we need be. Yet as DES consumer groups found out, inducing the federal government to monitor industry is difficult, because the political pressures on regulators can be overwhelming.

Several key uncertainties abounded in the DES research, and these foreshadow the uncertainties that haunt today's endocrine-disruptor policies. The significance of laboratory-animal experiments for people, the boundaries between synthetic and natural processes, the risks of low levels of exposure, and the significance of environmental influences on the developing fetus were all uncertain in the 1930s. They remain uncertain today, not because of lack of research effort but because of the complexity of endocrine systems. Using this complexity as a justification for continuing to expose people and environments to synthetic chemicals has proven to be a useful strategy for industry, but it is not one that is likely to protect Americans' health or the environment.

The regulatory agencies' willingness to approve DES was partly derived from the unwillingness of clinicians to pay heed to experimental evidence from laboratory animals. Karl John Karnaky, for example, insisted that the animal studies showing that DES caused fetal harm did not apply to people. In one publication, Karnaky noted that numerous lab studies had shown that DES was damaging to the fetus. And yet even after summarizing all the reports that DES harmed the fetus or prevented implantation of the fertilized egg, Karnaky went on to state that women are not laboratory animals, and thus there was no reason to believe that DES was harmful to women.[12]

For ethical and logistical reasons, it is impossible to gather experimental evidence on human subjects, but it is vital that we not delay regulatory action until human evidence is available. Nevertheless, when environmentalists and regulators attempt to limit human exposure because of findings from animal evidence, industry representatives argue that such precaution is absurd because rats are not people. Many doctors and lawmakers remain unwilling to believe that a compound's harmful effects on laboratory animals or wildlife can be extrapolated to humans. Frederick vom Saal noted in the 1990s that "before DES was used on five million women in the U.S., it was clear from animal studies that DES would be damaging to fetuses. But we have this absolutely bizarre notion that humans are separate from the rest of life on Earth. You will hear physician after physician say, 'But that's an animal. What relevance does that have to humans?' "[13] One clear message from the DES story is that we should not assume that research on laboratory animals has no meaning for people.

For decades, scientists and regulators debated the possible significance of low levels of exposure to synthetic chemicals. Even when researchers agreed that high levels of estrogens might cause harm, significant disputes remained about what those results might mean at the low levels common in the environment. Traditional toxicological models of risk posited dose-response models, where the dose makes the poison; in this model, low levels beneath a given threshold value would not be expected to cause harm. Industry advocates argued that these threshold values were based in sound science, but a careful reading of history reveals that they were often the result of political negotiation.

The boundary between natural and synthetic was also a continuing source of uncertainty. The drug companies argued that because bodies naturally produced estrogens, levels of additional estrogens that were just a fraction of the highest levels of the natural estrogens would not have a toxic effect. When Karnaky argued that DES treatment during pregnancy was safe, he pointed to the fact that a woman's body naturally produced high levels of estrogens during pregnancy, making the additional amounts from DES insignificant. Drug companies promoting DES manipulated the concept of naturalness, with its attendant implications of purity and safety. These same arguments remain potent today in debates over the safety of steroid hormones given to livestock.[14]

Another critical issue focused on the limits of technology and knowledge. If technology did not exist to measure a residue, did that mean the residue did not exist? If an effect could not be measured, was the effect therefore nonexistent? Industry initially argued that only effects and residues that were measurable existed. Scientists consulting with the FDA disputed this, arguing that an inability to detect liver damage from DES, for example, could mean that liver damage did not exist. But it might mean that available tests lacked the sensitivity to show slow, chronic changes. Initially, the FDA regulators agreed with this idea, refusing to assume that an inability to detect a residue or an effect meant the chemical was safe. Yet by 1947, this idea had been discarded, as the FDA joined the industry in arguing that if something could not be measured by available technologies, it did not exist.

Each time regulators reached the limits of their knowledge about the effects of a chemical exposure, they decided to move ahead and allow people to be exposed. Each time they vowed to use that new exposure as an experiment that would be monitored, so that policy makers could learn from the experiment. The toxic chemicals were released with the underlying assumption that "any major problems will emerge in good time for corrective action."[15] The corollary often cited was that if no major problems emerged, the compound must be safe. Yet when no monitoring is being done, that fundamental assumption is wrong. People may be dying in increased numbers from a particular chemical exposure, yet if their death rates are not being monitored, industry will continue to insist its products are safe. Time and again the federal agencies failed to learn from their own histories—sometimes because they lacked the funding and political power to insist on monitoring, and sometimes because they refused to pay attention to results.

The continuing failure of the FDA to regulate DES and the continuing insistence of physicians on prescribing the drug were closely linked to particular social constructions of diseases and treatments. As the medical historian Robert Bud argues about antibiotics, drugs "came to stand for the technical solution to infection, replacing control through prevention." Similarly, for many, DES represented a technical solution to menopause, then to miscarriage, and eventually to grain shortages. Advocates of progress tended to override concerns based in precaution. Rosner and Markowitz show how during Depression-era debates over the safety of lead paint, the lead industry "sought to co-opt the growing public health movement by identifying lead with modernity and health. . . . The themes of order, cleanliness, and purity that were hallmarks of the efforts to reform and sanitize American life were quickly incorporated into the promotional materials developed by the industry."[16] A similar pattern emerged for DES. Rather than addressing the larger ecological issues of "accidents of pregnancy," DES seemed to promise a technical solution that was cheap and, above all, modern. The pharmaceutical companies played upon these themes in their promotions of the synthetic hormone.

A crucial lesson from the DES history is that science alone cannot solve our chemical problems. Like many people trained in science, I had assumed that additional research on chemical and biological mechanisms would resolve policy conflicts. If we found more evidence about bisphenol A's effects on gene expression, then surely federal agencies would restrict the chemical. This has proven no more true than the hope in the 1960s that DES would be banned as soon as researchers linked DES use to cancer in women. With both chemicals, research findings alone did not lead to action. As the history of DES makes clear, the call for "more research!" has often become a way of delaying action, keeping profitable drugs and chemicals on the market as long as possible.[17]

In *The Secret History of the War on Cancer*, the epidemiologist Devra Davis describes how industry lobbyists have manipulated scientific uncertainty and risk-assessment protocols to delay action against toxic chemicals, particularly those that can cause cancer. Many industries have used a systematic approach to magnify doubt and delay regulatory action. "The first step is to feign blindness to a problem induced by a chemical by making sure that no records are kept concerning the health of the workers handling the chemical in question (without data, there is no proof). The second step is to create evidence systematically which refutes any possibility of a problem. Then sponsor carefully designed studies in low-risk populations which will confuse. Then fund yet more studies to suggest that doubt remains even after the dangers are clearly defined. Finally, use litigation, political lobbying and confidentiality clauses to delay publication of results for years or decades."[18] The DES histories show how successful these strategies can be.

The history of tobacco regulation offers a useful parallel. In *The Cigarette Century,* the public health historian Alan Brandt argues that the failure of U.S. doctors and regulators to respond to the dangers posed by tobacco was the result of a deliberate strategy to manufacture doubt. As Davis writes, "Working first with medical experts . . . the tobacco strategists counted on their ability to hire leading scientists who did not want to believe that smoking was harmful. With such an impressive front line, tobacco sympathizers carefully crafted doubt about what evidence is required before we can say that a given agent truly is a true threat to human health."[19] Strategies to promote pesticides and other endocrine disruptors were similar. In 1996, the Food Quality Protection Act (HR 1627) was passed, eviscerating the Delaney Clause and replacing it with the standard of "reasonable certainty that no harm will result from aggregate exposure to pesticide residue."[20] In other words, processed foods would be allowed to contain residues of carcinogenic pesticides. Zero tolerance was replaced with a risk-assessment standard that allows carcinogens to be present in processed foods if they create a "negligible risk" of causing cancer. This policy rested on the belief that scientists can indeed "assign accurate risks to the likelihood that a given quantity of a chemical will cause cancer."[21]

A group of toxicologists with the American Health Foundation lobbied particularly hard for the repeal of the Delaney Clause. One toxicologist, J. H. Weisburger, published a series of papers contending that the Delaney Clause was pointless because cancer was genetic, not environmental, in origin. In 1996, Weisburger argued that the Delaney Clause "was based on the hypothesis held in the 1950s that human cancers are due to environmental chemicals. This is clearly not true for the great majority of cancers and therefore, the Delaney Clause as framed has not saved any lives, is obsolete, and should be eliminated."[22] Weisburger and other staff at the American Health Foundation were key authors of a position statement on carcinogens in foods that was endorsed by the North American Society of Toxicologic Pathologists in 1995. In this position statement they argued that the Delaney Clause was essentially "irrational" because after the clause was passed in 1962, "major progress has been made in understanding mechanisms of cancer induction and in recognizing causes of human cancer. The Clause in conjunction with its present legal interpretation and implementation does not provide for rational, scientific evaluation of carcinogens." The position statement declared that animal studies were often inapplicable for humans, and chemicals in food should be monitored only on the basis of human, not animal, experiments.[23]

The American Health Foundation also worked under contract with tobacco companies, promoting research into "safe" cigarettes. We can trace abundant parallels between the tobacco lobbyists and the firms hired to slow regulation of global warming and toxic chemicals. Not only are the strategies the same; the people, funding sources, consultants, and public relations firms are often the same as well. The historian Naomi Oreskes argues that one key political tactic involves manufacturing a fake debate to undercut emerging scientific consensus. This tactic has been used against the consensus that sulfur and nitrogen emissions cause acid rain, the consensus that chlorofluorocarbons cause the hole in the ozone layer, the consensus that cigarette smoking causes cancer, the consensus on endocrine disruptors, and particularly the growing consensus on global warming. These efforts follow a similar pattern. First, deniers argue that the science is uncertain. Then they argue that the scientific concerns are exaggerated and the true risks are small, particularly compared to natural risks already existing in the environment. Finally, they state that technology will solve the problem, eliminating the need for government interference. The campaigns against environmental and public health regulation involve the same institutions, run by the same people, funded by the same sources.[24]

David Michaels and Celeste Monforton describe in detail the ways that the tobacco industry promoted scientific uncertainty to delay regulatory action. They write that "the tobacco industry recognized the value of magnifying the debate in the scientific community on the cause-and-effect relationship between smoking and lung cancer. In the 1960s, the Tobacco Institute published a journal entitled *Tobacco and Health Research,* aimed at physicians and scientists. The criteria for publishing articles in the journal were straightforward: 'The most important type of story is that which casts doubt on the cause-and-effect theory of disease and smoking.' In order to ensure that the message was clearly communicated, the public relations firm advised that headlines 'should strongly call out the point— Controversy! Contradiction! Other Factors! Unknowns!' "[25] While many people feel that this strategy has marked the tobacco industry as unique, Michaels and Monforton point out that there is nothing unusual about tobacco companies. The same firms are involved in the same activities today, particularly to block regulation of bisphenol A and other common chemicals that offer profits to some people and risks to millions.

The current Endocrine Disruptor Screening Program illustrates the problems with the argument that more science is the only rational solution. As part of the Food Quality Protection Act, Congress ordered the EPA to begin screening and testing chemicals and pesticides for endocrine-disrupting effects by 1999. The Endocrine Disruptor Screening Program would consist of a series of tests on laboratory animals designed to discover whether synthetic chemicals are endocrine disruptors. This sounded rational. With more than a hundred thousand chemicals on the market today, a coordinated screening program might help consumers and regulators know which chemicals interfere with hormones and which do not.

At first, enthusiasm was high within environmental and research communities. The journalists Susanne Rust, Meg Kissinger, and Cary Spivak write that "the EPA convened a committee of scientists from academia, the government and the chemical industry to lay the groundwork for testing these chemicals. They came up with a way to identify and test chemicals for the risks and get the information to the public. . . . Then-EPA administrator Carol Browner said in 1998 that her agency would begin fast-tracking efforts to screen these compounds by the end of that year. 'Some 15,000 chemicals used in thousands of common products, ranging from pesticides to plastics,' would be screened, Browner said." But a decade and nearly $80 million later, debates over protocols meant that not a single chemical had been screened.[26]

Delays and calls for more science are not the only ways the Endocrine Disruptor Screening Program has favored business as usual. Critics say that Stephen Johnson, the EPA administrator who served under the Bush administration from 2005 to 2009, favored industry in myriad ways, for example by allowing the Endocrine Disruptor Screening researchers to conduct lab tests that use a strain of rat that is essentially unresponsive to known hormone-disrupting chemicals.[27] In addition, the EPA allowed the study rats to be fed soy-based chow high in phytoestrogens that can mask the effect of endocrine-disrupting chemicals. The rats that were supposed to be controls could not be true controls after eating endocrine-disrupting phytoestrogens for dinner. The tests also used a dosage range that made it difficult to detect low-dose effects, which are the ones of greatest concern. And finally, the Endocrine Disruptor Screening Program did not require that tests be conducted on prenatal exposure, even though the fetus is the most vulnerable to exposure. "If your objective is not to find anything, that's the perfect way to do it," notes Frederick vom Saal.[28]

The reasons for most miscarriages, stillbirths, and premature births were a mystery during the DES era, and they remain a mystery today. Some miscarriages occur because of chromosomal abnormalities. Others come about because of hormonal problems such as diabetes, which was the problem for which DES was originally prescribed before its expansion into all cases of repeated miscarriage and then all cases of pregnancy. Many other "accidents of pregnancy" are intimately connected to what Sandra Steingraber calls the ecology of pregnancy: the larger web of relationships that interconnect a woman and her family with social, biophysical, and cultural environments.

Poverty, poor nutrition, stress, exposure to smoke, and exposure to environmental pollutants may all be involved in fetal death. Some of these involve individual choices that the mother makes, such as what she drinks, smokes, or eats. Historically, much of the emphasis on prenatal care has been on the mother's individual choices. Miscarriage is often portrayed as due to a flaw of the mother's. She had bad genes or she ate the wrong thing, she took a drink, or her body was inadequate for the task of fulfilling her female duty. As the authors of a 2004 health publication suggest, work to reduce the rate of low-birth-weight infants should focus on "improving maternal lifestyle choices."[29]

In 2003, a flurry of media attention greeted the publication of two studies that appeared to show that treatment with the hormone progesterone might reduce the risk of preterm babies.[30] Like the initial DES studies, these two studies were limited in extent, did not test to see whether the hormone reduced perinatal morbidity and mortality, and did not follow the babies into adulthood to find out whether the prenatal exposure affected their later health. As with DES, the mechanisms of the hormone's activity were unclear. The studies did not attempt to show whether progesterone was safe for fetal development, nor were they able to address the multitude causes that lead to preterm births — factors such as poverty, poor nutrition, air pollution, and obesity. A few researchers urged caution, pointing out the parallels to the DES history. Nevertheless, many interpreted the results just as an earlier generation had interpreted the early DES studies. One newspaper headline promised "A Shot of Hope" with progesterone. What the *New York Times* called "the toll of anguish" of premature labor might be solved with an inexpensive pill.[31]

The *New York Times* medical writer Jane Brody acknowledged that failed pregnancy is complicated: "Poverty, lack of prenatal care and chronic stress raise the risk as well. Women who smoke raise their risk by 20 percent to 30 percent. As many as 15 percent of all preterm births are attributable to smoking during pregnancy. Body weight also matters: the risk rises if a woman is too thin before pregnancy or gains too little or too much during the pregnancy. . . . Women with certain health problems are more likely to deliver prematurely. These include diabetes, whether it developed in pregnancy or beforehand; high blood pressure, either before or during pregnancy; and serious infections like bacterial pneumonia, kidney infection, acute appendicitis or a sexually transmitted disease."[32] But instead of addressing poverty, pollution, and racism, the media response raised hopes that with progesterone doctors could treat high-risk pregnancies with a pill. Yet as the DES tragedies illustrate, the ecology of a healthy pregnancy is intimately linked to the larger ecosystems, not just to the individual genes or choices a woman makes.

Industrial chemicals are abundant artifacts of a society that was brought into being within a highly specific cultural infrastructure against a deeper historical backdrop of evolution that occurred without their presence. And yet increasingly they are a part of the natural world, and as persistent chemicals, many of them will continue to be a part of our bodies far into the future.[33] Although we cannot remove the traces of these chemicals from our bodies or ecosystems, governments can decide to act with precaution.

Steingraber and the biologist Mary O'Brien offer alternative approaches to environmental regulation that recognize the need to make difficult decisions in the face of continuing uncertainty. Rather than precluding all action until everything is known about a chemical, these approaches advocate that anyone proposing to use a chemical should first compare a wide range of alternatives, then choose the least toxic alternative available for the given situation. Under the National Environmental Policy Act of 1970, federal land agencies are required to conduct a similar comparison of alternatives. Before creating a new forest-management plan, for example, the Forest Service must first provide an Environmental Impact Statement that describes the possible outcomes from a wide variety of reasonable alternatives. Theoretically, these may range from "do nothing" to "do everything." Comparing a wide range of alternatives can help decision makers realize that better approaches exist—particularly since the National Environmental Policy Act requires that federal land agencies involve the public in the planning process. As anybody connected with forest planning can tell you, the process is not perfect, but alternatives assessment does offer a way to bring diverse voices into the regulatory process and a way to search for the least toxic alternative.[34]

The most important lesson of the DES tragedy is the need for intelligent regulation to protect public health and the environment. Among the free market enthusiasts of the George W. Bush administration, regulation was a dirty word. During the eight years of anti-regulatory fervor between 2000 and 2008, environmental and health agencies were eviscerated. Political appointees chosen to head the agencies were often people who had spent their careers battling regulation rather than trying to make it more effective. Little wonder that the Food and Drug Administration and the Environmental Protection Agency are both in chaos. While it is impossible to say what changes President Obama's administration might bring to environmental policy, a respect for scientific evidence and a pragmatic understanding of the need for careful regulation and oversight should be among the highest priorities.

The financial chaos of fall 2008 illustrates the perils of ideology-driven deregulation. After years of deregulation and the abdication of oversight responsibilities, financial systems threatened to collapse. Advocates of deregulation had insisted that financial systems were resilient, able to absorb whatever was tossed at them, yet the chaos of 2008 finally made pundits rediscover the virtues of regulation. Why were so many powerful groups able to ignore the growing signs of instability and stress? As the historian Peter Perdue notes, "Governing elites resist looking too closely into historical roots of current crises; they suppress evidence and manipulate historical narratives to legitimate themselves. The fact that financial crises, or environmental crises, have reoccurred repeatedly even in recent memory doesn't guarantee that anyone will really want to address the fundamental causes. Historians have to recognize, and tell their readers, that impulses to denial, willful blindness, and ideological distortion are just as powerful as rational analysis in causing social change."[35] Historians must play exactly this role. We might not be able to prevent the powerful from willfully manipulating historical narratives to stay in power. But as the DES histories make clear, we can and should provide counternarratives that push back against these manipulations.

NOTES

Chapter 1. Disrupting Hormonal Signals

1. Sam Howe Verhovek, "Mysterious Force Attacks Small Western Tribe's Young in the Womb," *New York Times,* March 26, 2000.

2. The key work on endocrine disruptors intended for a general audience is Colborn, Dumanoski, and Myers, *Our Stolen Future.* Sheldon Krimsky offers a scholarly examination of the origins of the hypothesis in *Hormonal Chaos.* An early scientific review article summarizing effects of endocrine disruptors is Colborn, vom Saal, and Soto, "Developmental Effects." For science journalism, see Raloff's series for *Science News.* For the importance of bodies in environmental history, see Sellers, "Thoreau's Body."

3. Centers for Disease Control and Prevention, *Third National Report on Human Exposure.*

4. Vom Saal and Bronson, "Sexual Characteristics"; vom Saal, "Sexual Differentiation."

5. Krimsky, *Hormonal Chaos,* 4.

6. For overviews, see Raloff, "Gender Benders" and "Common Pollutants Undermine Masculinity." For selected research reports, see Angus et al., "Gonopodium Development," and Jenkins et al., "Identification of Androstenedione."

7. See Metcalfe, "Sex and Sewage," Guillette et al., "Developmental Abnormalities," and Nagler et al., "High Incidence."

8. Schettler, "Endocrine Disruptors." For imposex see LeBlanc, "Are Environmental Sentinels Signaling?" LeBlanc, "Steroid Hormone-Regulated Processes in Invertebrates," and Raloff, "Macho Waters."

9. For the CDC data showing a doubling in the incidence of severe hypospadias, see Paulozzi, Erickson, and Jackson, "Hypospadias Trends." For an analysis of international data that suggests the increases may have been leveling off in many countries since 1985, see Paulozzi, "International Trends." For an overview,

see Baskin, Himes, and Colborn, "Hypospadias and Endocrine Disruption." For cryptorchidism, see García-Rodríguez et al., "Exposure to Pesticides and Cryptorchidism." For testicular cancer, see McKiernan, Hensle, and Fisch, "Increasing Risk." See also Bergstrom et al., "Increase in Testicular Cancer Incidence," and Toppari et al., "Male Reproductive Health and Environmental Xenoestrogens."

The literature on declines in sperm counts is large and contentious. For findings supporting the hypothesis that sperm counts are declining, see Carlsen et al., "Evidence for Decreasing Quality of Semen," Swan, Elkin, and Fenster, "Have Sperm Densities Declined?" and Swan, Elkin, and Fenster, "Question of Declining Sperm Density Revisited." For findings suggesting that sperm counts are not declining, see Paulsen, Berman, and Wang, "Data from Men," and Fisch et al., "Semen Analyses in 1,283 Men." For prostate cancer, see Dinse et al., "Unexplained Increases in Cancer Incidence."

10. For a review of conflicting evidence on changes in the timing of puberty, see Zuckerman, "When Little Girls Become Women."

11. See Herman-Giddens, Slora, and Wasserman, "Secondary Sexual Characteristics," Kaplowitz et al., "Earlier Onset of Puberty in Girls," and Howdeshell, Hotchkiss, et al., "Exposure to Bisphenol A Advances Puberty."

12. McLachlan, "Environmental Signaling."

13. Myers, "Why Endocrine Disruption Challenges Current Approaches."

14. "Dose Makes the Poison."

15. Berkson, *Hormone Deception,* 51.

16. Eubanks, "Focus."

17. Ibid.

18. Cooper et al., "Atrazine Disrupts." Many other hormones influence gonadotropin-releasing hormone secretion as well, and feedback control over hormones is far more complex than this brief description can suggest.

19. McLachlan et al., "From Malformations to Molecular Mechanisms."

20. Berkson, *Hormone Deception,* 43.

21. Wake, "Integrative Biology." For an overview of epigenetics aimed at the general public, see Begley, "Second, Secret Genetic Code."

22. Newbold, "Perinatal Carcinogenesis"; Cunha et al., "New Approaches"; Newbold, Padilla-Banks, and Jefferson, "Adverse Effects"; Anway et al., "Epigenetic Transgenerational Actions." The study on females is Nilsson et al., "Transgenerational Epigenetic Effects." Fetal origins of disease are emerging as a key focus of research. Early studies on famine survivors found that when the mother went hungry during critical periods of fetal growth, the fetus survived by adapting cardiovascular, metabolic, and endocrine functions. These adaptations permanently change the function and structure of the body, leading to adult conditions such as obesity. See Szyf, "Dynamic Epigenome," and Myers, "Good Genes Gone Bad."

23. Spearow et al., "Genetic Variation in Susceptibility"; Gould, Schull, and Gorski, "DES Action in the Thymus"; Cook et al., "Interaction Between Genetic Susceptibility and Early-Life Environmental Exposure," citing King, Marks, and Mandell, "Breast and Ovarian Cancer Risks."

24. The research is in Bell, Hertz-Picciotto, and Beaumont, "A Case-Control Study." See Persky et al., "Effects of PCB Exposure," and Mendola et al., "Consumption of PCB-Contaminated Sport Fish." Sandra Steingraber in *Having Faith,* 143–149, describes the many effects PCBs can have on thyroid function.

25. Block et al., "In Utero Diethylstilbestrol"; Akbas, Song, and Taylor, "Hoxa10 Estrogen Response Element"; Fei, Chung, and Taylor, "Methoxychlor Disrupts"; Grossman, "Two Words." For bisphenol A risks, see Environment and Human Health, "Plastics." For miscarriages, see Sugiura-Ogasawara et al., "Exposure to Bisphenol A."

26. Foster et al., "Detection of Endocrine-Disrupting Chemicals."

27. Krimsky, *Hormonal Chaos,* 231.

28. Krimsky, "Epistemological Inquiry," citing Committee on Hormonally Active Agents in the Environment, *Hormonally Active Agents,* 15.

29. Myers, "Hormonally Active Agents"; Schettler, "Endocrine Disruptors."

30. Schettler, "Endocrine Disruptors," 232.

Chapter 2. Before World War II

1. McCoy, Reisch, and Tullo, "Facts and Figures of the Chemical Industry"; Chandler, *Shaping the Industrial Century.*

2. Dunlap, *DDT,* 19.

3. Riley, in USDA *Farmers' Bulletin* 7, cited in Whorton, *Before "Silent Spring,"* 71.

4. Ibid., 177; see also Dunlap, *DDT,* 5–6.

5. The green pigments of the great Victorian designer and socialist William Morris were among the most toxic wallpaper pigments. Morris was a shareholder and director in Devon Great Consols, then the biggest arsenic producer in the world, which his father had helped set up. Many of the company's workers died of arsenic-related illness, yet Morris continued to deny that arsenic could harm consumers. See Meharg, "Science in Culture."

6. Whorton, *Before "Silent Spring,"* 21–34, 40–45.

7. Ibid., 99.

8. Hilts, *Protecting America's Health,* 12–13.

9. Ibid., 13.

10. Whorton, *Before "Silent Spring,"* 109. In *The Politics of Purity,* Clayton Coppin and Jack High interpret Wiley's relations to industry differently, arguing that Wiley allied himself with business firms and used the Pure Food and Drug Act to further his own power. Coppin and High's emphasis on regulation as a competitive process is useful, but their argument that "the rules that govern market competition usually produce the result of serving the general good" (170) finds little support in the history of American chemical marketing.

11. Whorton, *Before "Silent Spring,"* 110. Markowitz and Rosner have an illuminating discussion of the history of the related concept of "threshold limit value" in their essay "Industry Challenges." Nash, "Purity and Danger"; Davis, "Unraveling the Complexities of Joint Toxicity."

12. See Mitman, Murphy, and Sellers, "Landscapes of Exposure," Nash, "Purity and Danger," and Nash, *Inescapable Ecologies.*

13. Nash, "Purity and Danger."

14. Dunlap, *DDT,* 54.

15. In their history of silicosis, the historians Gerald Markowitz and David Rosner show how contrasting views of labor and management shaped the development of the concept of the "threshold limit value," which would come to dominate industrial hygiene in the 1930s. The developers of threshold limit values were seeking to set standards that would protect the worker without burdening manufacturers. These standards appeared to be imposingly quantitative, yet they were the product of what Markowitz and Rosner call a "political compromise that traded on workers' health" ("Industry Challenges," 506). Nash notes that in developing threshold limit values, "modern toxicology normalized the problem of low-level chemical exposures, at least within the factory. . . . When toxicology moved outside the factory, it brought with it the assumption that industrial chemicals are a normal part of the environment and that the only relevant question to ask was at what level" ("Purity and Danger," 654). See also Sellers, *Hazards of the Job.*

16. Watkins, *Estrogen Elixir,* 26; Hilts, *Protecting America's Health,* 74.

17. Whorton, *Before "Silent Spring,"* 176–177.

18. Hilts, *Protecting America's Health,* 73

19. Watkins, *Estrogen Elixir,* 26.

20. Quoted in Whorton, *Before "Silent Spring,"* 175.

21. Tugwell, *Diary,* 85.

22. Whorton, *Before "Silent Spring,"* 201.

23. Appropriations Act of 1937, cited ibid., 200ff.; see also Dunlap, *DDT,* 47–52.

24. Hilts in *Protecting America's Health* explores the controversies over the regulatory power of the FDA, 72–107. Marks in *The Progress of Experiment* examines relations among researchers, industry, and regulators during the 1930s and 1940s.

Chapter 3. Help for Women over Forty

1. Helen Haberman, "Help for Women over Forty," *Reader's Digest* (November 1941). A copy of the article is in NARA, FDA, A1, Entry 5, General Subject Files, 1938–1974 (1941), Folder 526.1 November–December.

2. Ibid. Watkins in *Estrogen Elixir* has numerous examples of media depictions of menopause.

3. Oudshoorn, *Beyond the Natural Body,* 26, 39. See Watkins, *Estrogen Elixir,* for a summary of hormone research. For histories of sexual difference see Fausto-Sterling, *Sexing the Body,* Dreger, *Hermaphrodites,* Fausto-Sterling, *Myths of Gender,* Roughgarden, *Evolution's Rainbow,* and Rudacille, *Riddle of Gender.*

4. Meyerowitz, *How Sex Changed;* Oudshoorn, *Beyond the Natural Body.*

5. Bell, "Synthetic Compound Diethylstilbestrol." Robert Battey described his

procedure in 1880 in "Summary of the Results" in the *British Medical Journal.* Watkins discusses this operation in more detail in *The Estrogen Elixir,* 17.

6. As one researcher on DES, Dr. Edward Davis, stated in an interview with the FDA in 1940, the point of the drug was to control natural fluctuation in hormones (Dr. Ernest King, Medical Officer of the FDA, Memorandum of interview with Dr. M. Edward Davis, University of Chicago, November 1, 1940, NARA, FDA, A1, Entry 5, General Subject Files, 1938–1974 [1940], Folder 526.1-.11). Houcke's *Hot and Bothered* explores medical and cultural histories of menopause, while Watkins examines hormone-replacement therapy in *The Estrogen Elixir.*

7. Watkins, *Estrogen Elixir,* 16–18.

8. Tone and Watkins, *Medicating Modern America,* 64–70.

9. Bell, "Synthetic Compound Diethylstilbestrol"; Sze, "Boundaries and Border Wars." Scholarly works on the history of DES include Apfel and Fisher, *To Do No Harm,* and Dutton, *Worse Than the Disease.* Works for a general audience include Laitman Orenberg, *DES,* Fenichell and Charfoos, *Daughters at Risk,* Meyers, *DES,* and Seaman, *Greatest Experiment.* For a review of the scientific literature on DES effects see Giusti, Iwamoto, and Hatch, "Diethylstilbestrol Revisited."

10. The 1938 *Nature* paper is Dodds et al., "Oestrogenic Activity." Other studies from Dodds's laboratory include Dodds et al., "Synthetic Oestrogenic Compounds," and Dodds and Lawson, "Molecular Structure."

11. I thank an anonymous reviewer for *Environmental History* for suggesting this perspective.

12. These studies were included in the annotated bibliography prepared by Merck and Company in April 1941 and submitted to the FDA ("Stilbestrol [Diethylstilbestrol]: Annotated Bibliography" [hereafter "Stilbestrol Bibliography"] [Rahway, N.J.: Merck and Company, April 1941]).

13. Parkes, Dodds, and Noble, "Interruption of Early Pregnancy." See also Noble, "Functional Impairment." Both these articles are listed in "Stilbestrol Bibliography."

14. R. R. Greene et al. (1939) showed that DES modified embryonic sexual development in rats ("Stilbestrol Bibliography," 7). A French researcher, E. Wolff, showed that DES altered sexual development in chicken embryos (1939), while J. H. Gaarenstroom (1939) showed that when chicken eggs were injected with DES on the second day of brooding, all the hatchlings were female—no cocks were hatched. Albert Raynaud (1939) showed feminization of male embryos when their mothers were treated with DES, while R. R. Greene et al. (1940) showed that DES led to cryptorchidism in male rats and female development problems at puberty. All these studies are listed in "Stilbestrol Bibliography."

15. Cadbury, *Altering Eden,* postscript. See Dodds and Lawson, "Synthetic Oestrogenic Agents," and Dodds and Lawson, "Molecular Structure." In an FDA interview, one scientist mentioned the work of Dr. Reuben Gustafson, who "had observed carcinoma of the thyroid in 100 percent of the rats he had treated with

Stilbestrol" (E. King, Medical Officer of the FDA, Memorandum of interview with Dr. Melvin Boynton, Chicago October 29, 1940, NARA, FDA, A1, Entry 5, General Subject Files, 1938–1974 [1940], Folder 526.1-.13). Other researchers connected abnormal and continued uterine bleeding after DES treatment with an increased risk of uterine cancer (Small Committee Report files, 1941, FDA, FOIA. DES Microfiche no. 163, Folder 156 17/19).

16. Dr. Gordon A. Granger, Medical Officer of the FDA, Memorandum of interview with Dr. U. J. Salmon, July 13, 1940, NARA, FDA, A1, Entry 5, General Subject Files, 1938–1974 (1940), Folder 526.1-.11; Granger, Memorandum of interview with Dr. C. Kohn, Kansas City, July 26, 1940, NARA, FDA, A1, Entry 5, General Subject Files, 1938–1974 (1940), Folder 526.1-.11.

17. William Stoner of the Schering Corporation wrote to the FDA's James Durrett that while all estrogens were problematic, synthetic ones were likely to be even worse because they were novel substances in evolution: "It was stated that the mammalian organism has become accustomed to the action of certain hormones which may produce damage although they usually do not, while other substances having actions simulating those of the natural hormones, such as synthaline, dinitrophenol, stilbestrol, etc., more or less uniformly cause damage with which the organism has not learned to cope" (Stoner to Durrett, Letter of July 6, 1939, FDA, FOIA. DES Microfiche no. 166, Folder 159). See also Dr. Gordon A. Granger, Medical Officer of the FDA, Memorandum of interview with Dr. Samuel Sorkin, Chicago, August 1, 1940, NARA, FDA, A1, Entry 5, General Subject Files, 1938–1974 (1940), Folder 526.1-.11, and Granger, Memorandum of interview with Dr. C. P. Rhoads, New York, July 10, 1940, NARA, FDA, A1, Entry 5, General Subject Files, 1938–1974 (1940), Folder 526.1-.11.

18. "Estrogen Therapy—A Warning" (Editorial), *Journal of the American Medical Association* 113 (1939): 2312; Leech, "Stilbestrol: Preliminary Report"; Buxton and Engle, "Effects of the Therapeutic Use of Diethylstilbestrol."

19. "Synthetic Female Hormone Pills Considered Potential Danger." A copy is in NARA, FDA, A1, Entry 5, General Subject Files, 1938–1974 (1940), Folder 526.1-.13.

20. Dr. Raphael Kurzrock, "The Action of Diethylstilbestrol in Gynecological Dysfunctions" (typed paper), and accompanying letter from James Durrett, FDA, to Kurzrock, September 19, 1939, FDA, FOIA, DES Microfiche no. 166, Folder 159.

21. Ephraim Shorr to James Durrett, FDA, May 25, 1939, FDA, FOIA, DES Microfiche no. 166, Folder 159; Durrett, Memorandum of interview with Ephraim Shorr, M.D., Society of the New York Hospital, September 19 and 20, 1940, FDA, FOIA, DES Microfiche no. 166, Folder 159.

22. Shorr to Durrett, May 25, 1939; E. King, Medical Officer of the FDA, Memorandum of interview with Dr. L. M. Randall, Mayo Clinic, October 28, 1940, NARA, FDA, A1, Entry 5, General Subject Files, 1938–1974 (1940), Folder 526.1-.13; James Durrett, FDA, Memorandum of interview with Earl T. Engle, M.D., September 19 and 20, 1940, FDA, FOIA, DES Microfiche no. 166, Folder

159; Dr. Gordon A. Granger, Medical Officer of the FDA, Memorandum of interview with Dr. S. H. Geist, NYC, October 31, 1940, NARA, FDA, A1, Entry 5, General Subject Files, 1938–1974 (1940), Folder 526.1-.11.

23. Bernhard Zondek and Felix Sulman, "Inactivation of Diethylstilbestrol in the Organism," listed in "Stilbestrol Bibliography," 10; King, Memorandum of interview with Randall, October 28, 1940.

24. Memorandum to Dr. Herbert C. Calvery re: Literature on Stilbestrol, February 29, 1940, FDA, FOIA, DES Microfiche no. 166, Folder 159; William H. Stoner, Medical Research Division, Schering Corporation, to James Durrett, FDA, Letter of April 26, 1939, FDA, FOIA, DES Microfiche no. 166, Folder 159. Stoner sent Durrett four letters between April and May (April 26, April 28, May 5, May 25, ibid.) that listed eight studies showing toxic reactions from DES. "Stilbestrol Bibliography" lists some of the studies that showed fairly benign reactions (e.g., Varangot, February 4, 1939) but not those that showed more serious reactions (e.g., Varangot, March 1939), suggesting that in creating this bibliography, the compilers might have been selective about sifting the evidence.

25. Jack Curtis, FDA, Memorandum to Dr. Herbert C. Calvery, re: Literature on Stilbestrol, February 29, 1940, FDA, FOIA, DES Microfiche no. 166, Folder 159.

26. Kurzrock, "Action of Diethylstilbestrol," and accompanying letter from Durrett to Kurzrock, September 19, 1939; James Durrett, FDA, Memorandum of interview with Raphael Kurzrock, M.D., September 19 and 20, 1940, FDA, FOIA DES Microfiche no. 166, Folder 159.

27. Ephraim Shorr to James Durrett, FDA, September 15, 1939, and return letter from Durrett to Shorr, September 19, 1939, FDA, FOIA, DES Microfiche no. 166, Folder 159.

28. See Dr. Gordon A. Granger, Medical Officer of the FDA, Memorandum of interview with Dr. G. Taylor, Harvard, Boston, July 15, 1940, NARA, FDA, A1, Entry 5, General Subject Files, 1938–1974 (1940) Folder 526.1-.11; Granger, Memorandum of interview with Salmon, July 13, 1940; James Durrett, FDA, to Ephraim Shorr, May 3 1940, NARA, FDA, A1, Entry 5, General Subject Files, 1938–1974 (1940), Folder 526.1-.32.

29. Buxton and Engle, "Effects of the Therapeutic use of Diethylstilbestrol"; "Synthetic Female Hormone Pills Considered Potential Danger."

30. Watkins, *Estrogen Elixir*, 28.

31. R. W. Weilserstein of FDA, Memorandum of interview with Hans Lisser, I. Penchars, Allen Palmer, I. Perry, San Francisco, September 3, 1940, NARA, FDA, A1, Entry 5, General Subject Files, 1938–1974 (1940), Folder 526.1-.13.

32. Walter Campbell, Commissioner of FDA, to Mr. Charles Dunn, NYC, July 1940 (full date unreadable in microfilm), NARA, FDA, A1, Entry 5, General Subject Files, 1938–1974 (1940), Folder 526.1-.32.

33. Walter Campbell, Commissioner of FDA, to Merck & Company, re NDA 4076, November 3, 1941, NARA, FDA, A1, Entry 5, General Subject Files, 1938–1974 (1941), Folder 526.1, November–December.

34. Gillam and Bernstein, "Doing Harm."

35. Dr. Gordon A. Granger, Medical Officer of the FDA, Memorandum of interview with Dr. Harry M. Nelson, Detroit, January 19, 1940, NARA, FDA, A1, Entry 5, General Subject Files, 1938–1974 (1940), Folder 525.03–526; James Durrett, FDA, to Mrs. Agnes M. Sullivan, Elizabeth, N.J., September 16, 1940, NARA, FDA, A1, Entry 5, General Subject Files, 1938–1974 (1940), Folder 526.1-.32.

36. Dr. James W. Davis, Statesville, N.C., to Dr. James Durrett, FDA, February 7, 1941, NARA, FDA, A1, Entry 5, General Subject Files, 1938–1974 (1941), Folder 526.1, January to July; Ephraim Shorr to Dr R. P. Herrick and James Durrett, FDA, August 1941, NARA, FDA, A1, Entry 5, General Subject Files, 1938–1974 (1941), Folder 526.1, July–September. Susan Bell analyzes the contents of the case reports in "Gendered Medical Science."

37. On the revolving door problem see "The FDA's Cozy Little Relationship," http://www.goodhealthinfo.net/cancer/fda_cozy_relationship.htm (accessed August 18, 2008). On Klumpp, see Hilts, *Protecting America's Health,* 120.

38. Whorton, *Before "Silent Spring,"* 225.

39. Dr. Gordon A. Granger, Medical Officer of the FDA, Memorandum of interview with Dr. Robert T. Frank, New York, July 11, 1940, NARA, FDA, A1, Entry 5, General Subject Files, 1938–1974 (1940), Folder 526.1-.11. For a revealing summary of the endocreme case, see George Larrick Acting Chief, New Drug Division, FDA, Memorandum to Walter Campbell, Commissioner, FDA, October 1, 1941, NARA, FDA, A1, Entry 5, General Subject Files, 1938–1974 (1941), Folder 526.1, October. See also "Cancer Danger in Hormone Use," *Press-Scimitar,* Memphis, Tenn., May 20, 1940, NARA, FDA, A1, Entry 5, General Subject Files, 1938–1974 (1940), Folder 526.1-.13.

40. Watkins, *Estrogen Elixir,* 27.

41. "Proposed Lilly Circular on Stilbestrol," NDA 4041, FDA, FOIA, DES Microfiche no. 0004, Folder 3, December 26, 1941; Edgar B. Carter, Associate Director of Research of Abbott Laboratories, to Walton Van Winkle, FDA, regarding NDA 5442, November 4, 1944, FDA, FOIA, DES Microfiche no. 111, Folder 114; R. P. Herwick (FDA) to Mr. Edgar B. Carter of Abbott Laboratories, regarding NDA 5442, March 15, 1944.

42. Hunter Kennedy, Acting Chief, Drug Division, FDA, memorandum of interview with Dr. Joseph Rosin, representing Merck & Company, October 23, 1941, NARA, FDA, A1, Entry 5, General Subject Files, 1938–1974 (1941), Folder 526.1, November–December; Campbell to Merck & Company, November 3, 1941; for the revision, see Kennedy to Mr. R. L. Brault, Buffingtons, Inc., re: NDA 4423, January 26, 1942, NARA, FDA, A1, Entry 5, General Subject Files, 1938–1974 (1942), Folder 526.1, January–March.

43. Granger, Memorandum of interview with Sorkin, August 1, 1940.

44. P. B. Dunbar, Assistant Commissioner of Food and Drugs, to Mr. John S. Frosst of Charles Frosst & Company, regarding NDA 4061, December 3, 1941, FDA, FOIA, DES Microfiche no. 37, Folder 22/4.

45. E. King, Medical Officer of the FDA, Memorandum of interview with Dr. Cyril M. MacBryde, Saint Louis, October 30, 1940, NARA, FDA, A1, Entry 5, General Subject Files, 1938–1974 (1940), Folder 526.1-.13; Dr. Gordon A. Granger, Medical Officer of the FDA, Memorandum of interview with Dr. Kohn and Dr. W. W. Duke, Kansas City, Mo., August 5, 1940, NARA, FDA, A1, Entry 5, General Subject Files, 1938–1974 (1940), Folder 526.1-.11.

46. Granger, Memorandum of interview with Dr. Kohn, August 5, 1940.

47. H. Wales, Acting Chief, Interstate Division, FDA, to Mrs. Anna Greenzeig, Bronx, N.Y., March 21, 1942, NARA, FDA, A1, Entry 5, General Subject Files, 1938–1974 (1942a), Folder 526.1.-.10–526.3-.66.

Chapter 4. Bigger, Stronger Babies with Diethylstilbestrol

1. Hoover quoted in Lee and Howell, "Tall Girls," 1036.

2. Berkson, *Hormone Deception*, 62–63.

3. Board on Health Sciences Policy, *Preterm Birth;* Longnecker et al., "Association Between Maternal Serum Concentration of the Ddt Metabolite DDE."

4. W. R. M. Wharton, Chief, Eastern District, Memorandum to stations, Eastern District, June 25, 1942, NARA, FDA, A1, Entry 5, General Subject Files, 1938–1974 (1942a), Folder 526.1.-.10–526.3-.66.

5. Smith and Smith, "Prolan and Estrin"; White, "Diabetes Complicating Pregnancy"; White and Hunt, "Pregnancy"; Smith, Smith, and Hurwitz, "Increased Excretion"; Smith, "Diethylstilbestrol."

6. For criticisms see Hurwitz and Kuder, "Fetal and Neo-Natal Mortality," the comments of William Dieckmann in White, "Diabetes Complicating Pregnancy," 182, and the discussion in Gillam and Bernstein, "Doing Harm."

7. E. King, Medical Officer of the FDA, Letter to Robert A. Stormont, Medical Director, Abbott Laboratories, May 15, 1947, FDA, FOIA, DES Microfiche no. 20, Folder 11.

8. Karnaky cited in Gillam and Bernstein, "Doing Harm," 67. Karl John Karnaky to Dr. Newcomer of E. R. Squibb & Sons, 1946, FDA, FOIA, DES Microfiche no. 35, Folder 18/4/19. The letter offering to finance funeral costs is Karnaky to King, June 3, 1947, cited in Gilliam and Bernstein, "Doing Harm."

9. Karl John Karnaky, "Estrogenic Tolerance in Pregnant Women," *American Journal of Obstetrics and Gynecology* 51 (1947): 312–316, submitted as part of NDA 4047 by Abbott Laboratories for DES use in pregnancy, FDA, FOIA, DES Microfiche no. 20, Folder 11. The NDA package includes other key papers relating to DES use in pregnancy, including the Smiths' papers, the 1942 White and Hunt paper "Pregnancy Complicating Diabetes," and the White et al. paper "Prediction and Prevention of Late Pregnancy Accidents." "Application with Respect to Tablets of Diethylstilbestrol, 25 mg., by McNeil Laboratories, Inc.," December 15, 1947, FDA, FOIA, DES Microfiche no. 118, Folder 122, NDA 6394.

10. The material in this and the next two paragraphs is from H. Sidney Newcomer, Medical Department, E. R. Squibb and Sons, to R. P. Herwick, FDA, April

28, 1947, FDA, FOIA, DES Microfiche no. 26, Folder 17, NDA 4056. This file includes a series of testimonies from doctors who gave DES to their patients and reported on the outcomes to Dr. Newcomer. The doctors' and patients' names have been removed from the files.

11. Letter to H. Sidney Newcomer, November 19, 1946, ibid.

12. Granger objected to DES use in any pregnancy other than a diabetic pregnancy, writing that a company was "out of line in recommending [DES] for use in eclampsia, pre-eclampsia, and intrauterine fetal death. The first two conditions are too explosive to temporize with, and I do not see the evidence, other than in diabetes, to justify the recommendation," (Dr. Granger, FDA, Memorandum to E. King, Medical Officer of the FDA, regarding McNeil's NDA, no date, FDA, FOIA, DES Microfiche no. 118, Folder 122, NDA 6394).

13. Many of these studies were well enough known to be included in Burrow's 1945 *Biological Action of Sex Hormones.* See also Greene, "Embryology of Sexual Structure and Hermaphroditism," Kaplan, "Male Pseudohermaphroditism," Rosenblum and Melinkoff, "Preservation of the Threatened Pregnancy," and Lapa, "Diethylstilbestrol."

14. See Dieckmann et al., "Administration of Diethylstilbestrol During Pregnancy," and Swyer and Law, "Evaluation of the Prophylactic Ante-Natal Use of Stilboestrol."

15. Apfel and Fisher, *To Do No Harm.*

16. Gillam and Bernstein, "Doing Harm," 65–66.

17. Quoted in Hilts, *Protecting America's Health,* 118–119; ibid.

18. Seaman, *Greatest Experiment Ever Performed,* 5.

19. "Synthetic Female Hormone Pills Considered Potential Danger," 31.

20. Sze, "Boundaries and Border Wars," 795.

21. Gilliam and Bernstein, "Doing Harm," 68.

22. Dally, "Thalidomide." See also Cadbury, *Altering Eden,* 48.

23. Jane Maienschein, "Epigenesis and Preformationism," in *The Stanford Encyclopedia of Philosophy,* fall 2008 ed., http://plato.stanford.edu/archives/fall2008/entries/epigenesis (accessed May 28, 2009). For more on epigenetics and developmental biology, see Gilbert, "Genome in Its Ecological Context," and Nyhart, *Biology Takes Form.* By the Cold War, reaction in the West against the brutality of Soviet Lysenkoism had encouraged the rejection of ecological explanations in development. Because the Lysenkoists viewed the environment as crucial in shaping an organism's development, ecological developmental biologists in the West looked elsewhere for explanations.

Chapter 5. Modern Meat

1. Charles G. Durbin and Ralph W. Weilerstein, "The Food and Drug Administration, Poultry Feed Additives and Drugs," speech of August 3, 1960, NARA, FDA, A1, Entry 14, Office of Public Information Files, 1960–1964, Box 2.

2. Ibid.

3. Schell, *Modern Meat.*

4. Fitzgerald, *Every Farm a Factory;* Finlay, Review of Fitzgerald, 95.

5. Scott, *Seeing Like a State,* 4.

6. Navy pamphlet quoted in Bentley, *Eating for Victory,* 92–93; ibid., 111.

7. C. W. Crawford, Assistant Commissioner of Food and Drugs, FDA, to Mr. James A. Austin, Jensen-Salsbery Laboratories, Inc., Kansas City, Mo., May 8, 1944, NARA, FDA, A1, Entry 5, General Subject Files, 1938–1974 (1944a), Folder 526.1–526.11.

8. Harry J. Wick, Wick and Fry, Indianapolis, Ind., to FDA, March 20, 1945, and P. B. Dunbar to Mr. Harry J. Wick, Wick and Fry, Indianapolis, Ind., October 15, 1945, NARA, FDA, A1, Entry 5, General Subject Files, 1938–1974 (1945a), Folder 526.1.10.

9. Bentley, *Eating for Victory.*

10. H. E. Moskey, Chief, Veterinary Medical Section, FDA, to F. B Hutt, Professor of Animal Genetics, Cornell University, September 13, 1945, NARA, FDA, A1, Entry 5, General Subject Files, 1938–1974 (1945a), Folder 526.1.10; Leo Vogelman, Director, Research and Nutrition, Uncle Johnny Mills, to FDA, February 13, 1946; return letter from H. E. Moskey, Chief, Veterinary Medical Section, FDA, to Leo Vogelman, March 7, 1946; E. J. Ellerbusch, Manager, Holstein Produce and Hatchery, Holstein, Iowa, to FDA, January 25, 1946, all in NARA, FDA, A1, Entry 5, General Subject Files, 1938–1974 (1946), Folder 526.1. On October 8, 1945, the FDA office in Washington wrote to a midwestern branch ordering it to seize diethylstilbestrol pellets that had been made in a Kansas City laboratory for fattening and tenderizing poultry (P. B. Dunbar, Assistant Commissioner of Food and Drugs, to Central District Administration, re: Jensen-Salsbery Labs, Inc., October 8, 1945, NARA, FDA, A1, Entry 5, General Subject Files, 1938–1974 [1945a], Folder 526.1.10).

11. Memoranda from FDA staff for circulation, initials unreadable (WBW? and W?P), February 5, 1946, NARA, FDA. A1, Entry 5, General Subject Files, 1938–1974 (1946), Folder 526.1.

12. P. B. Dunbar, Assistant Commissioner of Food and Drugs, to Dr. David A. Bryce, Lederle Laboratories, Pearl River, N.Y., February 13, 1946, NARA, FDA, A1, Entry 5, General Subject Files, 1938–1974 (1946), Folder 526.1.

13. P. B. Dunbar, Assistant Commissioner of Food and Drugs, to Dr. David A. Bryce, Lederle Laboratories, Pearl River, N.Y., January 28, 1947, NARA, FDA, A1, Entry 5, General Subject Files, 1938–1974 (1947), Folder 526.1; P. H. E. Moskey, Chief, Veterinary Medical Section, FDA, Memorandum of interview with Director C. W. Sondern and various staff of White Laboratories, New Jersey, January 30, 1947, NARA, FDA, A1, Entry 5, General Subject Files, 1938–1974 (1947), Folder 526.1.

14. H. E. Moskey, Chief, Veterinary Medical Section, FDA, to E. I. Robertson, Associate Professor, Animal Husbandry, Cornell, July 21, 1947, NARA, FDA, A1, Entry 5, General Subject Files, 1938–1974 (1947), Folder 526.1.

15. L. I. Pugsley, Chief, Laboratory Service, Department of National Health and Welfare, Food and Drugs Division, Pharmacology Laboratory, Ottawa, to Dr. Bert J. Vos, Federal Security Agency, Food and Drugs Division, February 20, 1947, NARA, FDA, A1, Entry 5, General Subject Files, 1938–1974 (1947), Folder 526.1. Pugsley's letter was forwarded by the Food Security Agency to the FDA for a response.

16. Dr. Moskey, the chief of the medical division of the FDA, replied to the Canadians in Moskey to L. I. Pugsley, Chief, Laboratory Service, Department of National Health and Welfare, Ottawa, March 6, 1947, NARA, FDA, A1, Entry 5, General Subject Files, 1938–1974 (1947), Folder 526.1. See also FDA staffer, Letter to Dr. Bert J. Vos, Federal Security Agency, Food and Drugs Division, regarding the February 20, 1947, letter from Pugsley of Canada, February 27, 1947, NARA, FDA, A1, Entry 5, General Subject Files, 1938–1974 (1947), Folder 526.1.

17. J. S. Glover, Poultry Pathologist, Ontario Veterinary College, Canada, to Office of the FDA, July 28, 1947, NARA, FDA, A1, Entry 5, General Subject Files, 1938–1974 (1947), Folder 526.1.

18. H. E. Moskey, Chief, Veterinary Medical Section, FDA, to J. S. Glover, Poultry Pathologist, Ontario Veterinary College, Canada, August 6, 1947, and Glover to Moskey, August 8, 1947, NARA, FDA, A1, Entry 5, General Subject Files, 1938–1974 (1947), Folder 526.1.

19. H. E. Moskey, Chief, Veterinary Medical Section, FDA, to James A. Austin, Jensen-Salsbury Laboratories, September 10, 1947, NARA, FDA. A1, Entry 5, General Subject Files, 1938–1974 (1947) Folder 526.1.10.

20. C. W. Crawford, Associate Commissioner, FDA, to Geo. A. Montgomery, Associate Editor of *Capper's Farmer,* April 13, 1948, NARA, FDA, A1, Entry 5, General Subject Files, 1938–1974 (1948), Folder 526.1.

21. Gail Compton, "Pill Turns Rooster into Tender Soul: Female Hormone Does Trick," *Chicago Tribune,* February 1, 1948. This clipping was circulated and initialed by sixteen FDA staff, including H. E. Moskey, George Larrick, and P. B. Dunbar; see NARA FDA, A1, Entry 5, General Subject Files, 1938–1974 (1948), Folder 526.1.

22. Richard Waugh, Technical Director, Arapahoe Chemicals, Inc., to Dr. Robert Stormont, FDA, June 26, 1947, NARA, FDA. A1, Entry 5, General Subject Files, 1938–1974 (1947), Folder 526.1. The letter was circulated within the FDA for handwritten and initialed comments.

23. Dr. Robert Stormont, FDA, to Richard Waugh, Technical Director, Arapahoe Chemicals, Inc., July 9, 1947, NARA, FDA, A1, Entry 5, General Subject Files, 1938–1974 (1947), Folder 526.1. The letter was circulated within the FDA for handwritten and initialed comments.

24. Handwritten comments by FDA staff, ibid.

25. E. I. Robertson, Associate Professor, Animal Husbandry, Cornell, to H. E. Moskey, FDA, July 11, 1947, NARA, FDA, A1, Entry 5, General Subject Files,

1938–1974 (1947), Folder 526.1. Moskey replied, "It was the consensus of our medical staff that even if 15 mg pellets of diethylstilbestrol were to be accidentally swallowed by humans, there would be no harmful effect, and it was on this that we made the new drug applications effective. In this connection you may be interested in the two articles published by Dr. K. Y. Karnaky in the Southern Medical Journal" (Moskey to Robertson, July 21, 1947).

26. John H. Collins, FDA, "Drugs for Food-Producing Animals and Poultry Are a Problem," speech, specific audience not indicated, March 5, 1951, NARA, FDA, A1, Entry 10, Articles and Speeches, 1916–1964. Mink papers include Howell and Pickering, "Suspected Synthetic Oestrogen Poisoning," and Sundqvist, Amador, and Bartke, "Reproduction and Fertility in the Mink."

27. John H. Collins of the FDA wrote: "Since the initial product was introduced on the market in 1947 the practice of implanting poultry with diethylstilbestrol has grown rapidly to the extent that many millions of birds are treated annually. . . . Recent investigations have shown the directions for use are not being followed in all cases, with the result that pellet residues are very often found in the edible tissues. Examination of treated poultry when marketed disclosed that many of the treated birds contained pellet residues in the edible neck area. The Food and Drug Administration also developed evidence that partially unabsorbed pellets were present in some birds after they were completely dressed by butchers for home consumption" ("Use and Abuse of Diethylstilbestrol Pellets in Poultry," annotated draft of article intended for *American Egg and Poultry Review* 13 [September 1951], NARA, FDA, A1, Entry 10, Articles and Speeches, 1916–1964, Box 13). Nicholas Wade discusses the case in "DES."

28. "Hormones and Chickens"; Wade, "DES."

29. Marcus, *Cancer from Beef;* Raun and Preston, "History of Diethylstilbestrol Use in Cattle"; Marcus, *Cancer from Beef,* 78. See also Summons, "Animal Feed Additives," and Marcus, "The Newest Knowledge of Nutrition."

30. Raun and Preston were involved in agricultural hormone research. Their "History of Diethylstilbestrol Use" discusses this history in detail; much of their information is based on personal communications with Hale and Burroughs. Bennetts, Underwood, and Shier, "Specific Breeding Problem of Sheep"; Beck and Braden, "Studies on the Oestrogenic Substance"; Dohan et al., "Estrogenic Effects of Extracts"; Burroughs to Tilton, April 6, 1959, and April 24, 1959, quoted in Marcus, "Newest Knowledge of Nutrition," 75–76.

31. Raun and Preston, "History of Diethylstilbestrol Use"; Burroughs and Culbertson, "Effects of Trace Amounts of Diethylstilbestrol," 66, 67.

32. Raun and Preston, "History of Diethylstilbestrol Use," 3.

33. Hilts, *Protecting America's Health,* 120, citing his interviews with Rankin and Klumpp.

34. Marcus, *Cancer from Beef,* 22–25; Burroughs and Culbertston, "Effects of Trace Amounts of Diethylstilbestrol"; Burroughs et al., "Influence of Oral Administration"; Raun and Preston, "History of Diethylstilbestrol Use"; Wade, "DES."

35. Marcus, "Newest Knowledge of Nutrition," 66.

36. For the patent information, see Raun and Preston, "History of Diethylstilbestrol Use," citing T. W. Perry, personal communication; Marcus, *Cancer from Beef,* 25; Marcus, "Newest Knowledge of Nutrition," 67.

37. Dutton, *Worse Than the Disease,* 65.

38. Collins, "Drugs for Food-Producing Animals and Poultry Are a Problem."

39. "Growing with the 1960s," speeches delivered at the Tenth Conference, National Institute of Animal Agriculture, 1960, NARA, FDA, A1, Entry 14, Office of Public Information Files, 1960–1964, Box 2; Durbin and Weilerstein, "Food and Drug Administration."

40. Hadlow and Grimes, "Stilbestrol-Contaminated Feed," cited in Dutton, *Worse Than the Disease,* 61.

41. Granville Knight, W. Coda Martin, Rigoberto Iglesias, and William E. Smith, "Possible Cancer Hazard Presented by Feeding Diethystilbestrol to Cattle," paper presented at the Symposium on Medicated Feeds, 1956, with panel discussion, reprinted by the U.S. Department of Health, Education, and Welfare, October 1956, NARA, FDA, A1, Entry 10, Articles and Speeches, 1916–1964, Box 22, p. 168.

42. Ibid.

43. Ibid, 169, 168.

44. Ibid, 168.

45. Don Hines, response, ibid., 171.

46. Franz Gassner, response, ibid., 173.

47. U.S. Congress, "Hearings Before the House Select Committee to Investigate the Use of Chemicals in Food Products," House of Representatives, Eighty-first and Eighty-second Congress (Washington, D.C.: GPO, 1951); Dunlap, *DDT,* 67, 69. Ann Vileisis has an excellent discussion of consumer concerns and the House hearings in *Kitchen Literacy,* 160 ff.

48. Dunlap, *DDT,* 68.

49. Michaels and Monforton, "Manufacturing Uncertainty"; Vogel, "From 'the Dose Makes the Poison.'" Vogel quotes George Larrick's February 14, 1956, testimony before the House committee.

50. Nevis E. Cook, FDA, "The Food and Drug Administration and Feed Additives," speech delivered at the Poultry Health Conference, University of New Hampshire, February 4, 1960, NARA, FDA, A1, Entry 14, Office of Public Information Files, 1960–1964, Box 1.

51. Kleiner, "US May Relax Rules on Carcinogens in Food."

Chapter 6. Growing Concerns

1. "Growing with the 1960s," speeches delivered at the Tenth Conference, National Institute of Animal Agriculture, 1960, NARA, FDA, A1, Entry 14, Office of Public Information Files, 1960–1964, Box 2, p. 57.

2. Ibid, 7.

3. Vogel, "From 'the Dose Makes the Poison'"; Wargo, *Our Children's Toxic Legacy,* 106–107.

4. Russell, *War and Nature,* 156.

5. Ibid, 11.

6. Ibid., 14. See Lear, *Rachel Carson,* for an illuminating discussion of Carson.

7. Russell, *War and Nature,* 129.

8. Draize et al., "Summary of Toxicological Studies," and Woodard, Ofner, and Montgomery, "Accumulation of DDT," both cited in Russell, *War and Nature,* 157.

9. Russell, *War and Nature,* 150–163.

10. Ibid., 202; Cranor, "Some Legal Implications of the Precautionary Principle"; P. Daniel, *Toxic Drift;* Russell, *War and Nature,* 213–214.

11. Dunlap, *DDT,* 74.

12. Ibid.

13. Burlington and Lindeman, "Effect of DDT on Testes"; Welch et al., "Effect of Halogenated Hydrocarbon Insecticides"; Conney et al., "Effects of Pesticides"; newspaper headlines cited in Dunlap, *DDT,* 174.

14. Carson, *Silent Spring,* 121.

15. Sheldon Krimsky discusses McLachlan's emergence as a pivotal scientist in DES research in *Hormonal Chaos,* 11.

16. Silverman, "Schizophrenic Career of a 'Monster Drug,'" 404.

17. Ibid., 404, 405.

18. F. R. Davis, *Pesticides and Toxicology,* 15–49; Kelsey, "Historical Perspective"; Kelsey quoted in Seidman and Warren, "Frances Kelsey," 497.

19. Fisher, "Thalidomide"; Kelsey, "Historical Perspective."

20. Kelsey quoted in Seidman and Warren, "Frances Kelsey"; Fisher, "Thalidomide."

21. Kelsey, "Historical Perspective."

22. Seidman and Warren, "Frances Kelsey"; "Dr. Frances Kathleen Oldham Kelsey," *Changing the Face of Medicine: Celebrating America's Women Physicians,* exhibition, National Library of Medicine, online at http://www.nlm.nih.gov/changingthefaceofmedicine/physicians/biography_182.html (accessed November 26, 2008).

23. Fisher, "Thalidomide"; Seidman and Warren, "Frances Kelsey." The German manufacturer Chemie Grünenthal went on trial in 1968 for falsification of records and attempts to cover up the effects of thalidomide. Chemie Grünenthal eventually made an out-of-court agreement to pay 100 million marks into a trust fund for known victims in Germany. See Stephens and Brynner, *Dark Remedy,* for the legal battles over thalidomide in Europe.

24. Kelsey, "Historical Perspective."

25. Fisher, "Thalidomide"; Kelsey, "Historical Perspective."

26. In 1949, the *Journal of the American Medical Association* warned, "One must not lose sight of the fact that there is a possibility that large doses of female sex

hormones may affect a male fetus adversely." Medical reports described intersex conditions in children exposed to DES in utero; see Bongiovanni, DiGeorge, and Grumback, "Masculinization of the Female Infant"; and Waxman, Kelley, and Gartler, "Apparent Masculinization of Female Fetus." None of these articles is mentioned in FDA materials.

27. Dodds, "Contribution of Drug Research"; Dickens, "Edward Charles Dodds," 247.

28. Frances Kelsey, "Drugs in Pregnancy," speech given at the 111th Annual Meeting of the Minnesota State Medical Association, Rochester, Minn., May 19, 1964; Kelsey, "Chemical Teratogens," speech given at the Annual Health Conference, New York, June 11, 1963; Kelsey, "Problems Raised for the FDA by the Occurrence of Thalidomide Embryopathy in Germany, 1960–1961," speech given at the 91st Annual Meeting of the American Public Health Association, Kansas City, Missouri, November 14, 1963, all in NARA, FDA, A1, Entry 14, Office of Public Information Files, 1960–1961, Box 3, Kelsey.

29. See Kelsey, "Problems Raised for the FDA," Kelsey, "Drugs in Pregnancy," and Linda Bren, "Frances Oldham Kelsey: FDA Medical Reviewer Leaves Her Mark on History," *FDA Consumer Magazine* (March–April 2001).

30. Dutton, *Worse Than the Disease,* 65. Determining efficacy is not just a matter of rationally evaluating research findings. As the medical historians Andrea Tone and Elizabeth Siegel Watkins write, "The protocols and results of clinical trials may be best understood as negotiations among a wide set of social actors — regulators, patients, physicians, activists, and pharmaceutical executives — each of whom is differently invested in what gets counted as a 'medical fact'" (*Medicating Modern America,* 5).

31. Wade, "DES." Instead of altering its regulations or strengthening their enforcement, the USDA cut its residue-testing program in half in 1970, and the FDA allowed more than twice the amount of DES to be used in cattle.

32. "FDA Position on Diethylstilbestrol" (January 20, 1971), in "Regulation of Diethylstilbestrol (DES), 1972," hearing before the Subcommittee on Health of the Committee on Labor and Public Welfare, United States Senate, 92nd Congress, Second Session, on S. 2818, July 20, 1972 (Washington D.C.: U.S. Government Printing Office, 1972), 153.

33. Dutton, *Worse Than the Disease,* 68. The key study is Herbst, Ulfeder, and Poskanzer, "Adenocarcinoma of the Vagina."

34. The editor of the *New England Journal of Medicine* had sent Edwards a personal letter urging him to act; see U.S. Congress, House Committee on Government Operations, Subcommittee on Intergovernmental Regulations, chaired by L. H. Fountain, "Regulation of Diethylstilbestrol (DES): Its Use as a Drug for Humans and in Animal Feeds (Part 1), 92nd Congress, First Session, November 11, 1971 (Washington, D.C.: U.S. Government Printing Office, 1972), 1 [hereafter "Fountain hearings"]. Herbst testimony, Fountain hearings, 7.

35. Hollis Ingraham, New York Commissioner of Health, to FDA Commissioner Edwards, June 15, 1971, included in Fountain hearings, 15.

36. Herbst testimony, Fountain hearings, 4–6.

37. Editorial, *New England Journal of Medicine* 284 (1971): 878, regarding Herbst, Ulfeder and Poskanzer, "Adenocarcinoma of the Vagina"; discussion in Fountain hearings, 91–92; Greenwald testimony, Fountain hearings, 11.

38. Edwards testimony, Fountain hearings, 109–110.

39. Simmons testimony, Fountain hearings, 81.

40. Simmons testimony, Fountain hearings, 84; Goldberg testimony, Fountain hearings, 82.

41. Greenwald testimony, Fountain hearings, 16.

42. Edwards testimony, Fountain hearings, 51. Fountain asked Edwards directly whether he believed that Herbst and Greenwald's studies suggested that DES had been responsible for cancer in the women. Edwards minimized the findings.

43. Edwards testimony, Fountain hearings, 53; Simmons testimony, Fountain hearings, 92.

44. Edwards testimony, Fountain hearings, 53.

45. Hertz testimony, Fountain hearings, 58.

46. Ibid., 59.

47. Ibid., 70.

48. Ibid., 64. For Edwards's reluctance to regulate the tobacco industry, see for example, "Statement by Charles C. Edwards, M.D., Commissioner of Food and Drugs, Public Health Service, Department of Health, Education, and Welfare, Before the Subcommittee on the Consumer Committee on Commerce, United States Senate" (February 10, 1972), no. 1005111323/1355, Philip Morris Collection, Legacy Tobacco Documents Library, University of California, San Francisco, http://legacy.library.ucsf.edu/tid/kvv38e00 (accessed June 9, 2009). See Brandt, *Cigarette Century*, for an illuminating analysis of the federal agencies' failure to regulate tobacco.

49. Ibid., 68–69.

50. Ibid., 69.

51. Hertz, "Estrogen Problem," 1–2.

52. Krimsky, *Hormonal Chaos*, 11.

53. U.S. Congress, Senate Committee on Labor and Public Welfare, Subcommittee on Health, chaired by Edward Kennedy, hearings on "Regulation of Diethylstilbestrol (DES)," 92nd Congress, Second Session, July 20, 1972 (Washington, D.C.: U.S. Government Printing Office, 1972), 1–2 [hereafter "Kennedy hearings"].

54. Proxmire commentary, Kennedy hearings, 7–8.

55. Edwards and Simmons testimony, Kennedy hearings, 45 ff.

56. Friedman wrote that "estradiol is a natural hormone in man, and the metabolic pathways are already established so that in the event any is absorbed, it will be rapidly metabolized and excreted. DES, on the other hand, is a synthetic estrogen having ten times the potency of estradiol and is much more slowly metabolized and excreted, thus exerting its effects on target organs for a much longer

period of time" (Leo Friedman, Director of Toxicology, FDA, Memorandum to Dr. Virgil O. Wodicka, Director, Bureau of Foods, February 8, 1972, included in Kennedy hearings, 14–15.

57. Ibid.

58. Ibid.; Dr. M. Adrian Gross, FDA, Memorandum to Leo Friedman, Director, Division of Toxicology, and Dr. Daniel Banes, Director, Office of Pharmaceutical Research and Testing, Bureau of Drugs, FDA, December 5, 1971, included in Kennedy hearings, 124–130.

59. Friedman, Memorandum to Wodicka, February 8, 1972, 14–15; Nathan Mantel, Biometry Branch, National Cancer Institute, Memorandum to Dr. M. Adrian Gross, Bureau of Drugs, FDA, February 22, 1972, included in Kennedy hearings, 85–86.

60. Proxmire commentary, Kennedy hearings, 46.

61. Dutton, *Worse Than the Disease,* 72.

62. Edwards, quoted ibid., 78.

63. Allbright, Reifenstein, and Forbes, "Effect of Estrogen in Acromegaly"; Goldzieher, "Treatment of Excessive Growth." Goldzieher had long been a promoter of DES; as early as 1940, he admitted to "using large amounts of it" well before FDA approval. Moreover, he had a close relationship with estrogen manufacturers; see Dr. Gordon A. Granger, Medical Officer of the FDA, Memorandum of interview with Dr. M. A. Goldzieher, New York, July 11, 1940, NARA, FDA, A1, Entry 5, General Subject Files, 1938–1974 (1940), Folder 526.1-.11. See also Lee and Howell, "Tall Girls," 1036; Elliott, *Better Than Well,* 241–242.

64. Lee and Howell, "Tall Girls." See also Venn et al., "Oestrogen Treatment," Lever et al., "Tall Women's Satisfaction," and Louhiala, "How Tall Is Too Tall?" Women treated with estrogens in adolescence to suppress growth are more likely to have fertility problems (including 40 percent less chance of getting pregnant in any given menstrual cycle), increased time to pregnancy (the number of menstrual cycles required to conceive), and lower chances of a live birth. Nevertheless, as of 2007, one-third of American pediatric endocrinologists continued to offer growth-suppression hormonal treatments for tall girls.

65. Both quotations are in Dutton, *Worse Than the Disease,* 74.

66. In her review of Alan Marcus's *Cancer from Beef,* Apple writes, "Marcus is so intent on documenting the 'lost reverence' (p. 151) for experts and expertise that he does not address the important issue of how one should choose among competing claims to scientific and technological authority. Scientists who disagreed with [Wise] Burroughs [of Iowa State] are relegated to the margins of the scientific community; consumers who resisted DES-fed cattle are considered dupes of false science," 397.

67. Wade, "DES."

Chapter 7. Assessing New Risks

1. F. R. Davis, "Unraveling the Complexities of Joint Toxicity"; Cranor, *Toxic Torts.* In 1993 the Supreme Court of the United States set a legal precedent known as the *Daubert* standard, making toxic tort cases extremely difficult for plaintiffs to win. Under the *Daubert* standard, trial judges must determine whether the testimony of expert witnesses is admissible under the tests of relevance and reliability. In *Daubert v Merrell Dow Pharmaceuticals,* the plaintiffs claimed that their serious birth defects stemmed from drugs their mothers had taken while pregnant. To support their case, the plaintiffs relied on laboratory animal experiments, pharmacological analyses, and epidemiological evidence. But they could not provide proof that any single human exposure to the drug caused a particular birth defect, and the court ruled that the plaintiff's scientific testimony did not satisfy the burden of proof and was therefore inadmissible. *Daubert* has had an enormous effect on toxic torts because direct human evidence of harm from a particular exposure is rarely possible; see Berger, "What Has a Decade of *Daubert* Wrought?"

2. Silbergeld, "Risk Assessment," 8. During the 1970s, the federal government attempted to regulate pesticides and other chemical releases with one of three approaches: bans, technology-based control, and risk-benefit balancing. Technology-based control relied on new technologies such as scrubbers to reduce releases of the pollutants into the environment. Yet because biologically active levels of pollution could persist beneath detection limits, many environmentalists distrusted this approach. Persistent bioaccumulated chemicals were particularly resistant to technology-based controls, because measurements at the point of discharge could rarely predict or prevent long-term accumulation of toxics. Bans, exemplified by the Delaney Clause and its bans on carcinogens, prohibited production of a given chemical. Environmentalists often favored bans, but to toxicologists they were an insult to their profession, and industry liked them even less. The risk-benefit balancing approach, which industry generally favored, assumed that "there are levels of exposure where risks are very low or nonexistent such that the net benefits of continued use are clear" (ibid.).

3. Thornton, *Pandora's Poison,* 7; Batt, "Protecting Our Health."

4. Silbergeld, "Risk Assessment." The use of economics in the calculus of risk and benefit horrified many people who had been exposed to toxic chemicals, particularly daughters of mothers who had taken DES; these women demanded to know who had the right to decide that their suffering was justified by the economic benefit to industry.

5. Shubik, "Potential Carcinogenicity of Food Additives."

6. Silbergeld, "Risk Assessment"; Davis, *Secret History.*

7. Thornton, *Pandora's Poison,* 7.

8. Sheehan, "No-Threshold Dose-Response Curves"; vom Saal and Sheehan, "Challenging Risk Assessment"; vom Saal and Hughes, "Extensive New Litera-

ture Concerning Low-Dose Effects"; vom Saal and Welshons, "Large Effects from Small Exposures"; vom Saal, Nagel, et al., " Implications for Human Health."

9. Newbold et al., "Developmental Exposure."

10. Visser, *Cold, Clear, and Deadly,* 2; Harremoës et al., *Late Lessons,* 170–171. Robert N. Proctor, in *Cancer Wars,* offers a detailed critique of quantitative risk assessments. He writes that their effect has been "almost invariably to stymie health and environmental regulations. . . . Those who suffer the costs of pollution are often not the ones who reap the benefits. . . . Discussions of the costs of regulation too often fail to ask: Costs to whom? Benefits to whom?" (84). In *Pandora's Poison,* Joe Thornton notes that in risk assessment a lack of data is seen as evidence of safety, therefore "untested chemicals are allowed to be used without restriction. Since the vast majority of chemicals have not been subject to toxicity testing, ignorance becomes the dominant factor in environmental decisions" (7–8).

11. Hormones that are legal for use in American beef cattle as of 2008 include estradiol 17b, testosterone, progesterone, zeranol, trenbolone acetate, and melengestrol acetate.

12. Swan et al., "Semen Quality"; vom Saal, "Could Hormone Residues Be Involved?"

13. Jacobs, "U.S., Europe Lock Horns"; United States Mission to the European Union, "Primer on Beef Hormones."

14. Samuel Epstein, a professor of environmental and occupational medicine at the School of Public Health, University of Illinois Medical Center, filed an affidavit in support of the European Union ban. Epstein's figures are included in the European Commission on Health and Consumer Protection, "Opinion of the Scientific Committee on Veterinary Measures Relating to Public Health," and "Opinion of the Scientific Committee on Veterinary Measures Relating to Public Health: Review of Previous SCVPH Opinions." See also Epstein, "McDonald's Is Leading the Way."

15. Hanrahan, " RS20142."

16. Balter, "Scientific Cross-Claims Fly"; Cancer Prevention Coalition, "New Challenges on the Safety of U. S. Meat: Oprah Right for Other Reasons, Says Professor of Environmental Medicine at University of Illinois School of Public Health," press release, http://www.preventcancer.com/press/releases/feb2_98.htm (accessed July 26, 2008).

17. Cancer Prevention Coalition, "New Challenges on the Safety of U. S. Meat."

18. Ingersoll, "U.S. Launches Probe"; Jacobs, "U.S., Europe Lock Horns."

19. See Andersson and Skakkebaek, "Exposure to Exogenous Estrogens in Food."

20. Swan et al., "Semen Quality"; vom Saal, "Could Hormone Residues Be Involved?"; Cho et al., "Red Meat Intake"; Duby and Travis, "Influence of Dienestrol Diacetate."

21. Cho et al., "Red Meat Intake."

22. Perez Comas, "Precocious Sexual Development"; Fara et al., "Epidemic of Breast Enlargement."

23. Raloff, "Common Pollutants Undermine Masculinity."

24. Kolpin et al., "Pharmaceuticals, Hormones, and Other Organic Wastewater Contaminants"; Barnes et al., "Water-Quality Data."

25. Discussed in Raloff, "Hormones."

26. Orlando et al., "Endocrine-Disrupting Effects."

27. See Bhathena et al., "Effects of Omega-3 Fatty Acids"; Saldeen and Saldeen, "Omega-3 Fatty Acids"; Tanmahasamut et al., "Conjugated Linoleic Acid Blocks Estrogen Signaling"; Dhiman et al., "Conjugated Linoleic Acid Content of Milk"; Aro et al., "Inverse Association."

28. For an ecological analysis of grain-fed cattle versus pastured cattle, see the commentary by Jo Robinson and the articles collected at her Web site Eat Wild, http://www.eatwild.com/animals.html (accessed July 26, 2008), Schell, *Modern Meat,* and Pollan, *Omnivore's Dilemma*.

29. Meikle, *American Plastic.*

30. V. E. Yarsley and E. G. Couzens, "The Expanding Age of Plastics," quoted ibid., 69.

31. Hohn, "Moby-Duck," 45.

32. See Langston and Hillgarth, "Extent of Primary Molt," and Langston and Rohwer, "Molt/Breeding Tradeoffs."

33. One microgram is one one-millionth of a liter. Soto describes this discovery in her untitled autobiographical essay in *in-cites;* the research is reported in Soto et al., "p-Nonyl-phenol."

34. Vogel, "Politics of Plastics"; Grossman, "Two Words."

35. Dodds et al., "Oestrogenic Activity"; Dodds and Lawson, "Synthetic Oestrogenic Agents."

36. Factor, "Mechanisms"; Howdeshell, Peterman, et al., "Bisphenol A Is Released."

37. Olea et al., "Estrogenicity of Resin-Based Composites." A response can be found in Imai, "Comments." The response to this response is Olea, "Comments on." See also "Forum." Key studies on food-contact issues include Brotons et al., "Xenoestrogens Released," and Howe, Borodinsky, and Lyon, "Potential Exposure to Bisphenol A." Gross, "Toxic Origins of Disease," examines the question of safe fetal exposure to bisphenol A.

38. Vogel, "Battles over Bisphenol A."

39. Hunt et al., "Bisphenol A Exposure." For a perceptive discussion of Hunt's work, see Grossman, "Two Words." For the Japanese studies see Miyakoda et al., "Passage of Bisphenol A into the Fetus," Takahashi and Oishi, "Disposition of Orally Administered," and Josephson, "Breaching the Placenta."

40. vom Saal and Hughes, "Extensive New Literature Concerning Low-Dose Effects."

41. For more on the importance of positive controls in bisphenol A research, see vom Saal and Welshons, "Large Effects from Small Exposures."

42. Cunha et al., "New Approaches for Estimating Risk."

43. Hunt et al., "Bisphenol A Exposure"; Susiarjo et al., "Bisphenol A Exposure in Utero."

44. For popular reviews, see Grossman, "Two Words," and Gross, "Toxic Origins of Disease." For the research, see Newbold, Jefferson, and Banks, "Long-Term Adverse Effects," Baird and Newbold, "Prenatal Diethylstilbestrol (DES) Exposure," and Palmer et al., "Prenatal Diethylstilbestrol Exposure." Newbold and colleagues in 2007 found that low-dose exposure to bisphenol A by developing fetuses had effects that emerged in the adult, as is true of DES. For both chemicals, these effects were strongest not at the highest or the lowest doses but at intermediate dosages.

45. Newbold, quoted in Gross, "Toxic Origins of Disease."

46. Lang et al., "Association of Urinary Bisphenol A Concentration"; vom Saal and Myers, "Editorial: Bisphenol A and Risk of Metabolic Disorders."

47. Michaels and Monforton, "Manufacturing Uncertainty."

48. The quotation comes from a 1969 document from an executive at Brown & Williamson, a tobacco company now owned by R. J. Reynolds. The full quotation is in Michaels, "Doubt Is Their Product."

49. Nagel et al., "Relative Binding Affinity"; vom Saal quoted in Gross, "Toxic Origins of Disease." One typical "junk science" attack is Milloy, "Junk Science Report." Further research on bisphenol A and similarities to DES includes Markey et al., "In Utero Exposure to Bisphenol A."

50. Gupta, "Reproductive Malformation"; Elswick, Miller, and Welsch, "Comments."

51. The two key panels are vom Saal et al., "Chapel Hill Bisphenol A Expert Panel Consensus Statement," and National Toxicology Program, "CERHR Expert Panel Report for Bisphenol A." Critiques include Gross, "Toxic Origins of Disease," Vogel, "Battles over Bisphenol A," and Hileman, "Bisphenol A Vexations."

52. American Chemistry Council, press release, April 15, 2008, http://www.americanchemistry.com/s_acc/sec_news_article.asp?CID=206&DID=7240 (accessed July 26, 2008).

53. See, for example, John Dingell, Chairman of the House of Representatives Committee on Energy and Commerce, and Bart Stupak, Chairman of the Subcommittee on Oversight and Investigations, to J. N. Gerard, President and Chief Executive Officer, American Chemistry Council, April 2, 2008, Oversight and Investigations: The Public Record for the 110th Congress, http://energycommerce.house.gov/Press_110/110-ltr.040208.ACC.BPA.pdf (accessed June 2, 2009), and Needleman, "Case of Deborah Rice."

54. John Dingell, Chairman of the House of Representatives Committee on Energy and Commerce, and Bart Stupak, Chairman of the Subcommittee on Oversight and Investigations, to Stephen Johnson, Administrator, Environmental Protection Agency, March 13, 2008, archived at http://energycommerce.house.gov/Press_110/110-ltr.031308.EPA.BPA.pdf (accessed July 11, 2008). Dingell

and Stupak also noted that the dismissals of scientists who had worked on bisphenol A were troubling, for they seemed to "argue that scientific expertise with regard to a particular chemical and its human health effects is a basis for disqualification from a peer review board. This does not seem sensible on its face."

55. Dingell and Stupak to Gerard, April 2, 2008.

56. The Weinberg letter is P. Terence Gaffney, Vice President, Product Defense, the Weinberg Group. Inc., to Jane Brooks, Vice President, DuPont, April 29, 2003, EPA Docket EPA-HQ-OPPT-2003-0012. http://www.regulations.gov/fdmspublic/component/main?main=DocumentDetail&o=09000064800b9d37 (accessed June 2, 2009). The letter is discussed in Paul Thacker, "Weinberg Proposal," and David Roberts, "Uncovering the Weinberg Group." See also Michaels, "Doubt Is Their Product," Michaels and Monforton, "Manufacturing Uncertainty," and Michaels, "Scientific Evidence and Public Policy."

57. Environment Canada, "Screening Assessment"; "Dingell, Stupak Praise Nalgene and Wal-Mart Canada for Discontinuing Use of Bisphenol A," Committee on Energy and Commerce, April 17, 2008, http://energycommerce.house.gov/index.php?option=com_content&view=article&id=263&catid=17:benefits&Itemid=58 (accessed June 9, 2009); "With Children Still Exposed to Toxic Chemicals in Everyday Baby Products, Schumer Unveils New Legislation to Ban Bisphenol-A from All Children's Food and Beverage Containers," March 30, 2009, http://schumer.senate.gov/new_website/record.cfm?id=310796 (accessed June 2, 2009).

58. Silbergeld describes such pressures in "Risk Assessment."

Chapter 8. Sexual Development and a New Ecology of Health

1. Harder, "Boyish Brains," discussing Rubin et al., "Evidence of Altered Brain."

2. Sharpe and Skakkebaek, "Are Oestrogens Involved?"

3. Lou Guillette, interviewed on "Fooling with Nature," *Frontline*, PBS, June 2, 1998, transcript at http://www.pbs.org/wgbh/pages/frontline/shows/nature/etc/script.html (accessed July 26, 2008); "Masculinity at Risk"; Giwercman and Skakkebaek, "The Human Testis."

4. Although bodies can be conceived of, in Judith Butler's terms, as "a site of cultural inscription," they are also profoundly material (Butler, *Gender Trouble*, 164). Roberts, "Drowning in a Sea of Estrogens."

5. The historian of science and physician Vernon Rosario argues that "the notion that the Y chromosome determines male sex now appears to be grossly simplistic. . . . The genetic and molecular triggers for the complex steps in the embryonic development and differentiation of the reproductive system are emerging as multifactorial and highly interdependent. At multiple critical moments, various genes trigger other genes with an array of nonsexual functions in a dynamic play of shifting molecular signifiers." Sex determination is "embedded in a matrix of epidemiological, medical, historical, and social questions" ("Biology of Gender," 284). See also Fausto-Sterling, "Science Matters, Culture Matters."

6. Oudshoorn, *Beyond the Natural Body.*

7. Mittwoch, "Sex Determination in Mythology and History."

8. Fujimura, "Sex Genes," 59.

9. Institute of Medicine, *Exploring the Biological Contributions to Human Health,* 22–42.

10. Ibid., 5, 22, 33, 47.

11. Travis, "Modus Operandi." See also Gilbert, *Developmental Biology.* A flash animation of sexual development is available at "How Is Sex Determined?" *Nova Online,* http://www.pbs.org/wgbh/nova/miracle/dete_flash.html (accessed July 18, 2008).

12. Institute of Medicine, *Exploring the Biological Contributions to Human Health,* 50–72; Rea, *Chemical Sensitivity,* 1653; Raloff, "That Feminine Touch?"

13. This research is reviewed in Yucel et al., "Hypospadias." The effects of prenatal DES exposure are reviewed in Palmlund, "Exposure to a Xenoestrogen." Another useful review of DES effects on sexual development is Newbold, "Lessons Learned." Studies on effects on males include Fielden et al., "Gestational and Lactational Exposure," Toyama et al., "Neonatally Administered Diethylstilbestrol," and Yin, Lin, and Ma, "MSX2 Promotes Vaginal Epithelial Differentiation." In female mice, when DES blocks Wnt-7a gene expression, Wolffian ducts (precursors to adult reproductive structures that should develop in males and regress in females) regress normally, but Müllerian ducts (precursors to adult reproductive structures that should regress in males and develop in females) fail to develop properly, leading to extensive reproductive problems; see Mericskay, Carta, and Sasson, "Diethylstilbestrol Exposure in Utero," Ma and Sassoon, "PCBs Exert an Estrogenic Effect," Suzuki et al., "Gene Expression," and Newbold, Padilla-Banks, and Jefferson, "Adverse Effects."

14. Fausto-Sterling, *Myths of Gender,* 81; Fujimura, "Sex Genes."

15. Institute of Medicine, *Exploring the Biological Contributions to Human Health,* 54–55. Researchers working with the biologist David Sassoon discovered that the gene Wnt-7a plays a critical role in the development of both male and female reproductive tracts, modulating the regression of either the Müllerian or the Wolffian ducts. DES can suppress the activity of this gene; see Miller, Degenhardt, and Sassoon, "Fetal Exposure to DES," and Miller and Sassoon, "*Wnt-7a* Maintains Appropriate Uterine Patterning."

16. Colborn, Dumanoski, and Myers, *Our Stolen Future;* Joffe, "Infertility and Environmental Pollutants."

17. Singer, "Does Testosterone Build a Better Athlete?"

18. Lowe, "Testosterone, Carbon Isotopes, and Floyd Landis."

19. Crews, "Problem with Gender"; Fausto-Sterling, *Sexing the Body;* Roberts, "Biological Behavior?" Roberts, "Matter of Embodied Fact."

20. Weiss, "Sexually Dimorphic Nonreproductive Behaviors"; Joffe, "Infertility and Environmental Pollutants"; Newbold, "Gender-Related Behavior in Women."

21. Fausto-Sterling, *Sexing the Body.*

22. Ibid. See also Tierer, "Hormone Mistreatment."

23. Roberts, "Biological Behavior?" and Fausto-Sterling, "Bare Bones of Sex."

24. For the rates of intersex fish found in many British streams, see Jobling, Williams, et al., "Predicted Exposures to Steroid Estrogens." Susan Jobling's laboratory has published a series of key papers on steroids and sexual disruption in fish; see in particular Jobling, Coey, et al., "Wild Intersex Roach," and Jobling, Nolan, et al., "Widespread Sexual Disruption." Kerlin, "Presence of Gender Dysphoria."

25. See Fausto-Sterling, *Sexing the Body,* 53. Fausto-Sterling calculates that seventeen of every thousand human babies in the United States are currently born intersex rather than a typical male or female. Although two sexes are not enough to contain the range of human differentiation, American society has been anxious to eliminate intersexuality and put people into secure categories of male and female, usually through genital surgery on infants. Each year hundreds of infants are operated on to alter their genitalia, an increasingly controversial practice, as Fausto-Sterling describes.

26. Debates on toxic chemicals and possible links to both intersexuality and transgender conditions are discussed in Johnson, "Endocrine Disruptors and the Transgendered." See also Rudacille, *Riddle of Gender.* For an illuminating discussion of transgender issues, see Roughgarden, *Evolution's Rainbow.*

27. Fausto-Sterling, "Refashioning Race."

28. Roughgarden, *Evolution's Rainbow.*

29. Grosz, "Darwin and Feminism."

30. Roberts, "Drowning in a Sea of Estrogens"; Marler et al., "Role of Sex Steroids."

31. Langston et al., "Evolution of Body Size"; Langston, Rohwer, and Gori, "Experimental Analysis."

32. Fausto-Sterling, "Bare Bones of Sex."

33. Newbold, "Gender-Related Behavior in Women"; Rubin et al., "Evidence of Altered Brain Sexual Differentiation"; Richtera et al., "In Vivo Effects of Bisphenol A."

34. Zsarnovszky et al., "Ontogeny"; Steve Hentges, executive director of the American Plastics Council polycarbonate business unit, quoted in Barrett, "Endocrine Disruptors," A217.

35. Thornton, *Pandora's Poison,* 111–112.

36. Steingarber, *Having Faith;* Kroll-Smith and Lancaster, "Bodies, Environments," 204.

37. Weaver, "Epigenetic Programming."

38. Fox, "Tinkering with the Tinkerer," citing Bartholomew, "Interspecific Comparison," S4; Haraway, "Biopolitics of Postmodern Bodies," 219, 222.

39. Haraway, "Cyborg Manifesto," 164–165, 169.

40. Kroll-Smith and Lancaster, "Bodies, Environments," 204.

41. Tancrède, "Role of Human Microflora."

42. Allen, "Ecosystems and Immune Systems," 15. See also Forget and Lebel,

"Ecosystem Approach to Human Health"; Lebel, *Health;* Mergler, "Integrating Human Health"; Tannock, *Normal Microflora.*

43. Dubos, *Mirage of Health;* Thornton, *Pandora's Poison,* 117.

44. Schettler, "Endocrine Disruptors"; Thornton, *Pandora's Poison,* 58.

45. Kiesecker, "Synergism."

46. Skelly's research is not yet published, but it is described by Barringer in "Hermaphrodite Frogs Found."

47. Latour, *We Have Never Been Modern,* 1–2. For an illuminating discussion of these hybrids, see Fausto-Sterling, "Science Matters, Culture Matters."

48. Fausto-Sterling, "Science Matters, Culture Matters," 114.

Chapter 9. Precaution and the Lessons of History

1. "Wingspread Statement"; O'Brien, *Making Better Environmental Decisions.*

2. Boehmer-Christiansen, "Precautionary Principle in Germany."

3. Harremoës, Introduction, in Harremoës et al., *Late Lessons from Early Warnings,* 16.

4. Ibid.

5. Ibid., 14.

6. Markowitz and Rosner, "Industry Challenges," 502.

7. Ibid.

8. Neustadt and May, *Thinking in Time.*

9. Editorial team, "Twelve Late Lessons," in Harremoës et al., *Late Lessons from Early Warnings,* 168.

10. Ibid., 169; Scott, *Seeing Like a State,* 327–331, 343ff.

11. Editorial team, "Twelve Late Lessons," 169.

12. Karnaky, "Estrogenic Tolerance in Pregnant Women."

13. Vom Saal quoted in Berkson, *Hormone Deception,* 16.

14. "Proposed Lilly Circular on stilbestrol," NDA 4041, FDA, FOIA, DES Microfiche no. 0004, Folder 3, December 26, 1941.

15. Editorial team, "Twelve Late Lessons," 171.

16. Bud, "Antibiotics," 24; Markowitz and Rosner, "Industry Challenges," 504.

17. See Markowitz and Rosner, *Deceit and Denial.* For a discussion of growing corporate influences on epidemiology, see Pearce, "Corporate Influences," and Pearce, "Response."

18. Davis, *Secret History,* 40; Andreyev, "Very Vicious Circle."

19. Brandt, *Cigarette Century;* Davis, *Secret War,* 142.

20. Todd and Brown, "Pesticide Reform."

21. Kleiner, "US May Relax Rules."

22. Weisburger, "37 Year History of the Delaney Clause," 183.

23. Williams et al., "Risk Assessment of Carcinogens in Food," 209. See also Weisburger, "Does the Delaney Clause," and Weisburger, "Human Protection."

24. Naomi Oreskes, "The American Denial of Global Warming," online lecture

at http://www.uctv.tv/search-details.asp?showID=13459 (accessed November 29, 2008).

25. Michaels and Monforton, "Manufacturing Uncertainty," S40.

26. Rust, Kissinger, and Spivak, "Are Your Products Safe?"

27. Ambrose, "Scientists Criticize EPA Chemical Screening Program."

28. Vom Saal, quoted ibid.

29. Rosenberg, "Low Birth Weight."

30. The studies are da Fonseca et al., "Prophylactic Administration of Progesterone," and Meis et al., "Prevention of Recurrent Preterm Delivery."

31. Words of caution come from, among others, Michael Greene, "Progesterone and Preterm Delivery," and Bernstein, "Too Soon to Adopt Progesterone." Positive reports include "Progesterone Shots," and Rackl, "Shot of Hope"; Brody, "Preventing Premature Birth."

32. Brody, "Preventing Premature Birth."

33. Roberts and Langston, "Toxic Bodies/Toxic Environments."

34. Steingraber, *Living Downstream;* O'Brien, *Making Better Environmental Decisions.*

35. Quoted in "AHR Conversation."

BIBLIOGRAPHY

Archives

Individual documents from these archives are cited in the notes.

FDA, FOIA: Food and Drug Administration, Freedom of Information Office, 5600 Fisher Lane, Rockville, Md.

NARA, FDA: National Archives and Records Administration at College Park, Md., Record Group 88, Records of the Food and Drug Administration

Published Sources

To access DOI addresses, go to http://dx.doi.org/.

"AHR Conversation: Environmental Historians and Environmental Crisis." *American Historical Review* 113 (2008): 1431–1465. DOI: 10.1086/ahr.113.5.1431.

Akbas, G. E., J. Song, and H. S. Taylor. "A Hoxa10 Estrogen Response Element (Ere) Is Differentially Regulated by 17ss Estradiol and Diethylstilbestrol (DES)." *Journal of Molecular Biology* 340 (2004): 1013–1023.

Allbright, F., E. C. Reifenstein, and A. P. Forbes. "Effect of Estrogen in Acromegaly." In *Conference on Metabolic Aspects of Convalescence: Transactions of the 14th Meeting of Josiah Macy, Jr., Foundation.* New York: Josiah Macy, Jr., Foundation, 1946.

Allen, C. R. "Ecosystems and Immune Systems: Hierarchical Response Provides Resilience Against Invasions." *Conservation Ecology* 5 (2001): 15.

Ambrose, Sue Goetinck. "Scientists Criticize EPA Chemical Screening Program." *Dallas Morning News,* May 27, 2007.

Andersson, A. M., and N. E. Skakkebaek. "Exposure to Exogenous Estrogens in Food: Possible Impact on Human Development and Health." *European Journal of Endocrinology* 140 (1999): 477–485.

Andreyev, Jervoise. "A Very Vicious Circle." Review of Devra Davis, *The Secret*

History of the War on Cancer. Spectator, January 30, 2008. http://www.spectator.co.uk/print/the-magazine/books/476796/a-very-vicious-circle.thtml (accessed July 27, 2008).

Angus, R.A., H. McNatt, W. M. Howell, and S. D. Peoples. "Gonopodium Development in Normal A and 11-Ketotestosterone-Treated Mosquitofish (*Gambusia affinis*): A Quantitative Study Using Computer Image Analysis." *General and Comparative Endocrinology* 123 (2001): 222–234.

Anway, Matthew D., Andrea S. Cupp, Mehmet Uzumcu, and Michael K. Skinner. "Epigenetic Transgenerational Actions of Endocrine Disruptors and Male Fertility." *Science* 308 (2005): 1466–1469.

Apfel, Roberta J., and Susan M. Fisher. *To Do No Harm: DES and the Dilemmas of Modern Medicine.* New Haven: Yale University Press, 1984.

Apple, Rima. Review of Alan Marcus, *Cancer from Beef: DES, Federal Food Regulation, and Consumer Confidence. Technology and Culture* 37 (1996): 396–397.

Aro, A., S. Mannisto, I. Salminen, M. L. Ovaskainen, V. Kataja, and M. Uusitupa. "Inverse Association Between Dietary and Serum Conjugated Linoleic Acid and Risk of Breast Cancer in Postmenopausal Women." *Nutrition and Cancer* 38 (2000): 151–157.

Austen, Ian. "Bottle Maker to Stop Using Plastic Linked to Health Concerns." *New York Times,* April 18, 2008.

Bailey, Ronald. "Scared Senseless." *Wall Street Journal,* August 11, 2008.

Baird, D. D., and Retha Newbold. "Prenatal Diethylstilbestrol (DES) Exposure Is Associated with Uterine Leiomyoma Development." *Reproductive Toxicology* 20 (2005): 81–84.

Balter, Michael. "Scientific Cross-Claims Fly in Continuing Beef War." *Science* 284 (1999): 1453–1454.

Barnes, Kimberlee K., Dana W. Kolpin, Michael T. Meyer, E. Michael Thurman, Edward T. Furlong, Steven D. Zaugg, and Larry B. Barber. "Water-Quality Data for Pharmaceuticals, Hormones, and Other Organic Wastewater Contaminants in U.S. Streams, 1999–2000." U. S. Geological Survey Open-File Report 02-94. Iowa City: U.S. Geological Survey, 2002. http://toxics.usgs.gov/pubs/OFR-02-94/ (accessed June 6, 2009).

Barrett, Julia R. "Endocrine Disruptors: Bisphenol A and the Brain." *Environmental Health Perspectives* 114 (2006): A217.

Barringer, Felicity. "Hermaphrodite Frogs Found in Suburban Ponds." *New York Times,* April 8, 2008.

Bartholomew, G. A. "Interspecific Comparison as a Tool for Ecological Physiologists." In *New Directions in Ecological Physiology,* ed. M. E. Feder, A. F. Bennett, W. W. Burggren, and R. B. Huey, 11–37. New York: Cambridge University Press, 1987.

Baskin, Laurence S., Katherine Himes, and Theo Colborn. "Hypospadias and Endocrine Disruption: Is There a Connection?" *Environmental Health Perspectives* 109 (2001): 1175–1183.

Batt, Sharon. "Preventing Disease: Are Pills the Answer?" (2002). In *Protecting Our Health: New Debates: Women and Health Protection in Collaboration with "DES Action Canada."* http://www.whp-apsf.ca/en/documents/pills_prevent.html (accessed July 1, 2008).

Battey, Robert. "Summary of the Results of Fifteen Cases of Battey's Operation." *British Medical Journal* 1 (1880): 510–512.

Beck, A. B., and A. W. Braden. "Studies on the Oestrogenic Substance in Subterranean Clover: (*Trifolium subterraneum* L. var. *Dwalganup*)." *Australian Journal of Experimental Biology and Medical Science* 29 (1951): 273–279.

Begley, Sharon. "How a Second, Secret Genetic Code Turns Genes On and Off." *Wall Street Journal*, July 23, 2004.

Bell, Erin M., Irva Hertz-Picciotto, and James J. Beaumont. "A Case-Control Study of Pesticides and Fetal Death Due to Congenital Anomalies." *Epidemiology* 12 (2001): 148–156.

Bell, Susan. "Gendered Medical Science: Producing a Drug for Women." *Feminist Studies* 21 (1995): 469–500.

———. "The Synthetic Compound Diethylstilbestrol (DES), 1938–1941: The Social Construction of a Medical Treatment." Ph.D. diss. Brandeis University, 1981.

Bennetts, H. W., E. J. Underwood, and F. L. Shier. "A Specific Breeding Problem of Sheep on Subterranean Clover Pastures in Western Australia." *Australian Veterinary Journal* 22 (1946): 10–11.

Bentley, Amy. *Eating for Victory: Food Rationing and the Politics of Domesticity.* Champaign: University of Illinois Press, 1998.

Berger, Margaret. "What Has a Decade of Daubert Wrought?" *American Journal of Public Health* 95, suppl. 1 (2005): S59–S65.

Bergstrom, R., H.-O. Adami, M. Mohner, W. Zatonski, H. Storm, Anders Ekhom, S. Tretli, L. Teppo, O. Akre, and T. Hakulinen. "Increase in Testicular Cancer Incidence in Six European Countries." *Journal of the National Cancer Institute* 88 (1996): 727–733.

Berkson, D. Lindsey. *Hormone Deception: How Everyday Foods and Products Are Disrupting Your Hormones—and How to Protect Yourself and Your Family.* Chicago: Contemporary Books, 2000.

Bernstein, Peter S. "Too Soon to Adopt Progesterone for the Prevention of Preterm Delivery." *Medscape Ob/Gyn and Women's Health* 8 (2003): 1.

Bhathena, S. J., E. Berlin, J. T. Judd, Y. C. Kim, J. S. Law, H. N. Bhagavan, R. Ballard-Barbash, and P. P. Nair. "Effects of Omega-3 Fatty Acids and Vitamin E on Hormones Involved in Carbohydrate and Lipid Metabolism in Men." *American Journal of Clinical Nutrition* 54 (1991): 684–688.

Bland, J. "Conception and Development." In *About Gender: Conception and Development* (1998). http://www.gender.org.uk/about/04embryo/44_cncp.htm (accessed July 21, 2008).

Block, K., A. Kardana, P. Igarashi, and H. S. Taylor. "In Utero Diethylstilbestrol

(DES) Exposure Alters Hox Gene Expression in the Developing Müllerian System." *FASEB Journal* 14 (2000): 1101–1108.

Board on Health Sciences Policy, Institute of Medicine of the National Academies, *Preterm Birth: Causes, Consequences, and Prevention*. Washington, D.C.: NAS Press, 2006. http://www.iom.edu/CMS/3740/25471/35813/35975.aspx (accessed July 22, 2008).

Boehmer-Christiansen, Sonja. "The Precautionary Principle in Germany—Enabling Government." In *Interpreting the Precautionary Principle,* ed. Tim O'Riordan and James Cameron, 31–61. London: Earthscan, 1994.

Bongiovanni, A. M., A. M. DiGeorge, and M. M. Grumback. "Masculinization of the Female Infant Associated with Estrogenic Therapy Alone During Gestation: Four Cases." *Journal of Clinical Endocrinology* 19 (1959): 1004–1011

Brandt, Alan. *The Cigarette Century: The Rise, Fall, and Deadly Persistence of the Product That Defined America*. New York: Basic, 2007.

Breast Cancer Fund. *State of the Evidence: The Connection Between Breast Cancer and the Environment*. 5th ed. Ed. Janet Gray. San Francisco: Breast Cancer Fund, 2008.

Brody, Jane. "Preventing Premature Birth and Its Toll of Anguish." *New York Times,* April 8, 2003.

Brotons, J. A., M. F. Olea-Serrano, M. Villalobos, V. Pedraza, and N. Olea. "Xenoestrogens Released from Lacquer Coatings in Food Cans." *Environmental Health Perspectives* 103 (1995): 608–612.

Bud, Robert. "Antibiotics: From Germophobia to the Carefree Life and Back Again: The Lifecycle of the Antiobiotic Brand." In *Medicating Modern America: Prescription Drugs in History,* ed. Andrea Tone and Elizabeth Siegel Watkins, 17–41. New York: New York University Press, 2007.

Burlington, Howard, and Verlus F. Lindeman. "Effect of DDT on Testes and Secondary Sex Characteristics of White Leghorn Cockerels." *Society for Experimental Biology and Medicine Proceedings* 74 (1950): 48–51.

Burroughs, Wise, and C. C. Culbertson. "The Effects of Trace Amounts of Diethylstilbestrol in Rations of Fattening Steers." *Science* 120 (1954): 66–67.

Burroughs, Wise, C. C. Culbertson, Edmund Cheng, W. H. Hale, and Paul Homeyer. "The Influence of Oral Administration of Diethylstilbestrol to Beef Cattle." *Journal of Animal Science* 14 (1955): 1015–1024.

Burrows, Harold. *Biological Action of Sex Hormones*. Cambridge: Cambridge University Press, 1945.

Butler, Judith. *Gender Trouble: Feminism and the Subversion of Identity*. 2nd ed. New York, Routledge, 1999.

Buxton, C. L. and Earl T. Engle. "Effects of the Therapeutic Use of Diethylstilbestrol." *Journal of the American Medical Association* 113 (1939): 2318–2320.

Cadbury, Deborah. *Altering Eden: The Feminization of Nature*. New York: St. Martin's, 1999.

Carlsen, E., A. Giwercman, N. Keiding, and N. Skakkebaek. "Evidence for Decreasing Quality of Semen During Past 50 Years." *British Medical Journal* 305 (1992): 609–613.

Carson, Rachel. *Silent Spring.* New York: Houghton Mifflin, 1962.

Centers for Disease Control and Prevention. *National Report on Human Exposure to Environmental Chemicals.* Atlanta, Ga.: Centers for Disease Control, 2001.

——. *Third National Report on Human Exposure to Environmental Chemicals.* Atlanta, Ga.: Centers for Disease Control, 2005.

Chandler, Alfred D. *Shaping the Industrial Century: The Remarkable Story of the Evolution of the Modern Chemical and Pharmaceutical Industries.* Cambridge: Harvard University Press, 2005.

Cho, Eunyoung, Wendy Y. Chen, David J. Hunter, Meir J. Stampfer, Graham A. Colditz, Susan E. Hankinson, and Walter C. Willett. "Red Meat Intake and Risk of Breast Cancer Among Premenopausal Women." *Archives of Internal Medicine* 166 (2006): 2253–2259.

Colborn, T., F. S. vom Saal, and A. M. Soto. "Developmental Effects of Endocrine-Disrupting Chemicals in Wildlife and Humans." *Environmental Health Perspectives* 101 (1993): 378–384.

Colborn, Theo, Dianne Dumanoski, and John Peterson Myers. *Our Stolen Future: Are We Threatening Our Fertility, Intelligence, and Survival?* New York: Plume, 1997.

Committee on Hormonally Active Agents in the Environment, National Research Council. *Hormonally Active Agents in the Environment.* Washington, D.C.: National Academies Press, 1999. http://books.nap.edu/catalog.php?record_id=6029 (accessed July 21, 2008).

Conney, A. H., R. M. Welch, R. Kuntzman, and J. J. Burns. "Effects of Pesticides on Drug and Steroid Metabolism." *Pharmacology and Therapeutics* 8 (1966): 2–8.

Cook, J. D., B. J. Davis, S. Cai, J. C. Barrett, C. J. Conti, and C. L. Walker. "Interaction Between Genetic Susceptibility and Early-Life Environmental Exposure Determines Tumor-Suppressor-Gene Penetrance." *Proceedings of the National Academy of Sciences* 2005 (102): 8644–8649.

Cooper, Ralph L., Tammy E. Stoker, Lee Tyrey, Jerome M. Goldman, and W. Keith McElroy. "Atrazine Disrupts the Hypothalamic Control of Pituitary-Ovarian Function." *Toxicological Sciences* 53 (2000): 297–307.

Coppin, Clayton A., and Jack High. *The Politics of Purity: Harvey Washington Wiley and the Origins of Federal Food Policy.* Ann Arbor: University of Michigan Press, 1999.

Cranor, Carl. "Some Legal Implications of the Precautionary Principle: Improving Information-Generation and Legal Protections." *Human and Ecological Risk Assessment* 11 (2005): 29–52.

——. *Toxic Torts: Science, Law, and the Possibility of Justice.* Cambridge: Cambridge University Press, 2006.

Crawford, J. D. "Treatment of Tall Girls With Estrogen." *Pediatrics* 62 (1978): 1189–1195.

Crews, David. "The Problem with Gender." *Psychobiology* 16 (1998): 321–334.

Crews, David, Judith Bergeron, and John McLachlan. "The Role of Estrogen in Turtle Sex Determination and the Effect of PCBs." *Environmental Health Perspectives* 103, suppl. 7 (1995): 73–77.

Cunha, Gerald R., John-Gunnar Forsberg, Robert Golden, Arthur Haney, Taisen Iguchi, Retha Newbold, Shanna Swan, and Wade Welshons. "New Approaches for Estimating Risk from Exposure to Diethylstilbestrol." *Environmental Health Perspectives* 107, suppl. 4 (1999): 625–630.

Da Fonseca, E. B., R. E. Bittar, M. H. Carvalho, M. Zugaib. "Prophylactic Administration of Progesterone by Vaginal Suppository to Reduce the Incidence of Spontaneous Preterm Birth in Women at Increased Risk: A Randomized Placebo-Controlled Double-Blind Study." *American Journal of Obstetrics and Gynecology* 188 (2003): 419–424.

Dally, Ann. "Thalidomide: Was the Tragedy Preventable?" *Lancet* 351 (1998): 1197–1199

Daniel, Pete. *Toxic Drift: Pesticides and Health in the Post–World War II South.* Baton Rouge: University of Louisiana Press / Washington, D.C.: Smithsonian Museum of Natural History, 2005.

Davis, Devra. *The Secret History of the War on Cancer.* New York: Basic, 2007.

Davis, Devra Lee, Pamela Webster, Hillary Stainthorpe, Janice Chilton, Lovell Jones, and Rikuo Doi. "Declines in Sex Ratio at Birth and Fetal Deaths in Japan, and in U.S. Whites but Not African Americans." *Environmental Health Perspectives* 115 (2007): 941–946.

Davis, Frederick Rowe. "Pesticides and Toxicology: Episodes in the Evolution of Environmental Risk Assessment (1937–1997)." Ph.D. diss. Yale University, 2001.

———. "Unraveling the Complexities of Joint Toxicity of Multiple Chemicals at the Tox Lab and the FDA." *Environmental History* 13 (2008): 674–683.

Davis, J. R., R. C. Brownson, and R. Garcia. "Family Pesticide Use Suspected of Causing Child Cancers." *Archives of Environmental Contamination Toxicology* 24 (1993): 87–92.

Dhiman, T. R., G. R. Anand, L. D. Satter, and M. W. Pariza. "Conjugated Linoleic Acid Content of Milk from Cows Fed Different Diets." *Journal of Dairy Science* 82 (1999): 2146–2156.

Dickens, F. "Edward Charles Dodds." *Biographical Memoirs of Fellows of the Royal Society* 21 (1975): 227–267.

Dieckmann, W. J., M. E. Davis, L. M. Rynkiewicz, and R. E. Pottinger. "Does the Administration of Diethylstilbestrol During Pregnancy Have Therapeutic Value?" *American Journal of Obstetrics and Gynecology* 66 (1953): 1062–1075.

Dinse, G. E., D. M. Umbach, A. J. Sasco, D. G. Hoel, and D. L. Davis. "Unexplained Increases in Cancer Incidence in the United States from 1975 to 1994." *Annual Review of Public Health* 20 (1996): 173–209.

Dodds, E. C. "The Contribution of Drug Research to Science and Society." In

TheDevelopment and Control of New Drug Products, ed. M. Pernarowski and Marvin Darrach, 1–7. Vancouver, B.C.: Evergreen, 1972.

Dodds, E. C., L. Goldberg, W. Lawson, and R. Robinson. "Oestrogenic Activity of Certain Synthetic Compounds." *Nature* 141 (1938): 247–249.

———. "Synthetic Oestrogenic Compounds Related to Stilbene and Diphenylethane. Part 1." *Proceedings of the Royal Society of London, Series B: Biological Sciences* 127 (1939): 140–167.

Dodds, E. C., and W. Lawson. "Molecular Structure in Relation to Oestrogenic Activity. Compounds Without a Phenanthrene Nucleus." *Proceedings of the Royal Society. London, Series B: Biological Sciences* 125 (1938): 222–232.

———. "Synthetic Oestrogenic Agents Without the Phenanthrene Nucleus." *Nature* 137 (1936): 996.

Dohan, F. C., E. M. Richardson, R. C. Stribley, and P. Gyorgy. "The Estrogenic Effects of Extracts of Spring Rye Grass and Clover." *Journal of the American Veterinary Medical Association* 118 (1951): 323–324.

"Dose Makes the Poison." In *Chemical Safety in the Laboratory.* Yale Office of Environmental Health and Safety, http://learn.caim.yale.edu/chemsafe/references/dose.html (accessed January. 21, 2009).

Draize, John H., Geoffrey Woodard, O. Garth Fitzhugh, Arthur A. Nelson, R. Blackwell Smith, Jr., and Herbert O. Calvery. "Summary of Toxicological Studies of the Insecticide DDT." *Chemical and Engineering News* 22 (1944): 1503–1504.

Dreger, Alice. *Hermaphrodites and the Medical Invention of Sex.* Cambridge: Harvard University Press, 1998.

Dubner, Stephen J. "Freakonomics: Is There a 'Secret History of the War on Cancer'? Ask for Yourself." *New York Times,* November 1, 2007. http://freakonomics.blogs.nytimes.com/2007/11/01/is-there-a-secret-history-of-the-war-on-cancer-ask-for-yourself/ (accessed July 27, 2008).

Dubos, René. *Mirage of Health: Utopias, Progress, and Biological Change.* New York: Harper and Brothers, 1959.

Duby, R. T., and H. F. Travis. "Influence of Dienestrol Diacetate on Reproductive Performance of Female Mink (*Mustela vision*)." *American Journal of Veterinary Research* 32 (1971): 1599–1602.

Dunlap, Thomas. *DDT: Scientists, Citizens, and Public Policy.* Princeton: Princeton University Press, 1981.

Dutton, Diana. *Worse Than the Disease: Pitfalls of Medical Progress.* Cambridge: Cambridge University Press, 1988.

Elliott, Carl. *Better Than Well: American Medicine Meets the American Dream.* New York: Norton, 2003.

Elswick, Barbara A., Frederick J. Miller and Frank Welsch. "Comments to the Editor Concerning the Paper Entitled 'Reproductive Malformation of the Male Offspring Following Maternal Exposure to Estrogenic Chemicals' by C. Gupta." *Experimental Biology and Medicine* 226 (2001): 74–75.

Environment and Human Health, Inc.: John Wargo, Susan S. Addiss, Nancy O. Alderman, D. Barry Boyd, Russell L. Brenneman, David R. Browh, Mark R. Cullen, Robert G.LaCamera, and Hugh S. Taylor. "Plastics That May Be Harmful to Children and Reproductive Health." North Haven, Conn.: Environmental and Human Health, June 2008. http://www.ehhi.org/reports/plastics/ehhi_plastics_report_2008.pdf (accessed July 21, 2008).

Environment Canada. "Screening Assessment for the Challenge: Phenol, 4,4′-(1-methylethylidene)bis- (Bisphenol A)." Chemical Abstracts Service Registry No. 80-05-7. October 2008. http://www.ec.gc.ca/substances/ese/eng/challenge/batch2/batch2_80-05-7.cfm (accessed June 6, 2009).

Epstein, Samuel S. "McDonald's Is Leading the Way, but Hasn't Gone Far Enough Yet" *AScribe: The Public Interest Newswire,* July 1, 1999. http://www.rightlivelihood.org/fileadmin/Files/PDF/Literature_Recipients/Epstein/Epstein_-_McDonald_s.pdf_.pdf (accessed July 26, 2008).

"Estrogen Therapy—A Warning." Editorial. *Journal of the American Medical Association* 113 (1939): 2312.

Eubanks, Mary W. "Focus: Hormones and Health." *Environmental Health Perspectives* 105 (1997): 5.

European Commission on Health and Consumer Protection. "Opinion of the Scientific Committee on Veterinary Measures Relating to Public Health: Assessment of Potential Risks to Human Health from Hormone Residues in Bovine Meat and Meat Products," April 30, 1999, http://ec.europa.eu/food/fs/sc/scv/out21_en.pdf (accessed July 26, 2008).

———. "Opinion of the Scientific Committee on Veterinary Measures Relating to Public Health: Review of Previous SCVPH Opinions of 30 April 1999 and 3 May 2000 on the Potential Risks to Human Health from Hormone Residues in Bovine Meat and Meat Products," http://ec.europa.eu/food/fs/sc/scv/out50_en.pdf (accessed July 26, 2008).

Factor, A. "Mechanisms of Thermal and Photodegradations of Bisphenol A Polycarbonate." In *Polymer Durability: Degradation, Stabilization, and Lifetime Prediction,* ed. R. L. Clough, N. C. Billingham, and K. T. Gillen, 59–76. Washington, D.C.: American Chemistry Society, 1996.

Fara, G. M., G. Del Corvo, S. Bernuzzi, A. Bigatello, C. Di Pietro, S. Scaglioni, and G. Chiumello. "Epidemic of Breast Enlargement in an Italian School." *Lancet* 8137 (1979): 295–297.

Fausto-Sterling, Anne. "The Bare Bones of Sex. Part 1: Sex and Gender." *Signs: Journal of Women in Culture and Society* 30 (2005): 1491–1527.

———. *Myths of Gender: Biological Theories About Women and Men.* New York: Basic, 1992.

———. "Refashioning Race: DNA and the Politics of Health Care." *Differences: A Journal of Feminist Cultural Studies* 15 (2004): 1–37.

———. "Science Matters, Culture Matters." *Perspectives in Biology and Medicine* 46 (2003): 109–124.

——. *Sexing the Body: Gender Politics and the Construction of Sexuality.* New York: Basic, 2000.

Fei, X., H. Chung, and H. S. Taylor. "Methoxychlor Disrupts Uterine Hoxa10 Gene Expression." *Endocrinology* 146 (2005): 3445–3451.

Fenichell, Stephen, and Lawrence S. Charfoos. *Daughters at Risk: A Personal DES History.* New York: Doubleday, 1981.

Fielden, Mark R., Robert G. Halgren, Cora J. Fong, Christophe Staub, Larry Johnson, Karen Chou, and Tim R. Zacharewski. "Gestational and Lactational Exposure of Male Mice to Diethylstilbestrol Causes Long-Term Effects on the Testis, Sperm Fertilizing Ability in Vitro, and Testicular Gene Expression." *Endocrinology* 143 (2002): 3044–3059

Finlay, Mark. Review of Deborah Fitzgerald, *Every Farm a Factory. History* 32 (2004): 95.

Fisch, H., E. T. Goluboff, A. H. Olson, J. Feldshuh, S. J. Broder, and D. H. Barad. "Semen Analyses in 1,283 Men from the United States over a 25-Year Period: No Decline in Quality." *Fertility and Sterility* 65 (1996): 1009–1014.

Fisher, Jack C., M.D. "Thalidomide: Tragedy and Regulatory Response." American Council on Science and Health (February 24, 2004). http://www.acsh.org/healthissues/newsID.618/healthissue_detail.asp (accessed June 9, 2009).

Fitzgerald, Deborah. *Every Farm a Factory: The Industrial Ideal in American Agriculture.* New Haven: Yale University Press, 2003.

Forget, G., and J. Lebel. "An Ecosystem Approach to Human Health." *International Journal of Occupational and Environmental Health* 7 (2001): S3–S38.

"Forum." *Environmental Health Perspectives* 104, no. 4 (1996). http://www.ehponline.org/docs/1996/104-4/forum.html#dental.

Foster, W., S. Chan, L. Platt, and C. Hughes. "Detection of Endocrine-Disrupting Chemicals in Samples of Second Trimester Human Amniotic Fluid." *Journal of Clinical Endocrinology and Metabolism* 85 (2000): 2954–2957.

Fox, Glen. "Tinkering with the Tinkerer: Pollution Versus Evolution." *Environmental Health Perspectives* 103 (1995): S4.

Fujimura, Joan. "Sex Genes: A Critical Sociomaterial Approach to the Politics and Molecular Genetics of Sex Determination." *Signs: Journal of Women in Culture and Society* 32 (2006): 49–82.

García-Rodríguez, R., M. García-Martín, M. Nogueras-Ocaña, J. de Dios Luna-del-Castillo, M. Espigares García, N. Olea, and P. Lardelli-Claret. "Exposure to Pesticides and Cryptorchidism: Geographical Evidence of a Possible Association." *Environmental Health Perspectives* 104 (1996): 1090–1095.

Gilbert, Scott F. *Developmental Biology.* 6th ed. Sunderland, Mass.: Sinauer Associates, 2000. http://www.ncbi.nlm.nih.gov/books/bv.fcgi?rid=dbio (accessed July 18, 2008).

——. "The Genome in Its Ecological Context: Philosophical Perspectives on Interspecies Epigenesis." *Annals of the New York Academy of Sciences* 981 (2002): 202–218.

Gillam, Richard, and Barton J. Bernstein. "Doing Harm: The DES Tragedy and Modern American Medicine." *Public Historian* 9 (1987): 57–82.

Giusti, R. M., K. Iwamoto, and E. E. Hatch. "Diethylstilbestrol Revisited: A Review of the Long-Term Health Effects." *Annals of Internal Medicine* 122 (1995): 778–788.

Giwercman, A., and N. E. Skakkebaek. "The Human Testis—An Organ at Risk?" *International Journal of Andrology* 15 (1992): 373–375.

Goetinck Ambrose, Sue. "Scientists Criticize EPA Chemical Screening Program." *Dallas Morning News,* May 27, 2007. http://www.dallasnews.com/sharedcontent/dws/news/healthscience/stories/052707dnentendocrine.3a08215.html (accessed July 27, 2008).

Goldzieher, M.A. "Treatment of Excessive Growth in the Adolescent Female." *Journal of Clinical Endocrinology and Metabolism* 16 (1956): 249–252.

Gould, K. A., J. D. Schull, and J. Gorski. "DES Action in the Thymus: Inhibition of Cell Proliferation and Genetic Variation." *Molecular and Cellular Epidemiology* 170 (2000): 31–39.

Gray, E. L., Jr., J. Ostby, R. L. Cooper, and W. R. Kelce. "The Estrogenic and Antiandrogenic Pesticide Methoxychlor Alters the Reproductive Tract and Behavior Without Affecting Pituitary Size or LH and Prolactin Secretion in Male Rats." *Toxicology and Industrial Health* 15 (1999): 37.

Gray, E. L., Jr., J. Ostby, E. Monosson, and W. R. Kelce. "Environmental Antiandrogens: Low Doses of the Fungicide Vinclozolin Alter Sexual Differentiation of the Male Rat." *Toxicology and Industrial Health* 15 (1999): 48.

Gray, E. L., Jr., C. Wolf, C. Lambright, P. Mann, M. Price, R. L. Cooper, and J. Ostby. "Administration of Potentially Antiandrogenic Pesticides (Procymidone, Linuron, Iprodione, Chlozolinate, p, p′-DDE, and Ketoconazole) and Toxic Substances (Dibutyl- and Diethylhexyl Phthalate, PCB 169, and Ethane Dimethane Sulphonate) During Sexual Differentiation Produces Diverse Profiles of Reproductive Malformations in the Male Rat." *Toxicology and Industrial Health* 15 (1999): 94.

Greene, Michael. "Editorial: Progesterone and Preterm Delivery—Déjà Vu All Over Again." *New England Journal of Medicine* 348 (2003): 2453.

Greene, R. R. "Embryology of Sexual Structure and Hermaphroditism." *Journal of Clinical Endocrinology* 4 (1944): 335–348.

Gross, Liza. "The Toxic Origins of Disease." *PLoS Biology* 5 (2007): e193 doi:10.1371/journal.pbio.0050193.

Grossman, Elizabeth. "Two Words: Bad Plastic." *Salon* (August 2, 2007). http://www.salon.com/news/feature/2007/08/02/bisphenol/index1.html (accessed July 21, 2008).

Grosz, Elizabeth. "Darwin and Feminism: Preliminary Investigations for a Possible Alliance." *Australian Feminist Studies* 14 (1999): 31–45.

Guillette, L. J., Jr., T. S. Gross, G. R. Masson, J. M. Matter, H. F. Percival, and A. R. Woodward. "Developmental Abnormalitiies of the Gonad and Abnormal Sex Hormone Concentrations in Juvenile Alligators from Contaminated and Control Lakes in Florida. *Environmental Health Perspectives.*" 102 (1994): 680–688.

Guillette, Louis J., Jr., and D. Andrew Crain, eds. *Environmental Endocrine Disrupters: An Evolutionary Perspective.* London: Taylor and Francis, 2000.

Gupta, Chandra. "Reproductive Malformation of the Male Offspring Following Maternal Exposure to Estrogenic Chemicals." *Proceedings of the Society for Experimental Biology and Medicine* 224 (2000): 61–68.

Hadlow, W. J., and Edward F. Grimes. "Stilbestrol-Contaminated Feed and Reproductive Disturbances in Mice." *Science* 122 (1955): 643–644.

Hanrahan, Charles E. "RS20142: The European Union's Ban on Hormone-Treated Meat." Congressional Research Service Report for Congress, National Library for the Environment, December 19, 2000. http://ncseonline.org/NLE/CRSreports/agriculture/ag-63.cfm (accessed June 2, 2009).

Haraway, Donna. "The Biopolitics of Postmodern Bodies: Constitutions of Self in Immune System Discourse." In *American Feminist Thought at Century's End: A Reader,* ed. Linda S. Kauffman, 199–233. Cambridge: Blackwell, 1994.

———. "A Cyborg Manifesto: Science, Technology, and Socialist-Feminism in the Late Twentieth Century." In her *Simians, Cyborgs and Women: The Reinvention of Nature,* 149–181. New York: Routledge, 1991.

Harder, B. "Boyish Brains." *Science News* 169 (2006): 276–277.

Harremoës, Poul, David Gee, Malcolm MacGarvin, Andy Stirling, Jane Keys, Brian Wynne, and Sofia Guedes Vaz, eds. *Late Lessons from Early Warnings: The Precautionary Principle, 1896–2000.* European Environment Agency Environmental Issue Report 22. Luxembourg: Office for Official Publications of the European Communities, 2001.

Harries, J. E., D. A. Sheahan, S. Jobling, P. Matthiessen, P. Neall, J. P. Sumpter, T. Tylor, and N. Zaman. "Estrogenic Activity in Five United Kingdom Rivers Detected by Measurement of Vitellogenesis in Caged Male Trout." *Environmental Toxicology and Chemistry* 16 (1997): 534–542.

Henricks D. M., S. L. Gray, J. J. Owenby, and B. R. Lackey. "Residues from Anabolic Preparations After Good Veterinary Practice." *APMIS* 109 (2001): 273–283.

Herbst, A. L., H. Ulfelder, and D. C. Poskanzer "Adenocarcinoma of the Vagina: Association of Maternal Stilbestrol Therapy with Tumor Appearance in Young Women." *New England Journal of Medicine* 284 (1971): 878–881.

Herman-Giddens, Marcia E., E. J. Slora, and R. C. Wasserman. "Secondary Sexual Characteristics and Menses in Young Girls Seen in Office Practice: A Study from the Pediatric Research in Office Settings Network." *Pediatrics* 99 (1997): 505–512.

Hertz, Roy. "The Estrogen Problem—Retrospect and Prospect." In *Estrogens in the Environment II: Influences on Development,* ed. John A. McLachlan, 1–11. New York: Elsevier Science, 1985.

Hileman, Bette. "Bisphenol A Vexations." *Chemical and Engineering News* 85 (2007): 31–33.

Hilts, Philip J. *Protecting America's Health: The FDA, Business, and 100 Years of Regulation*. New York, Knopf, 2003.

Hohn, Donovan. "Moby-Duck." *Harper's* (January 2007): 39–62.

"Hormones and Chickens." *Time* (December 21, 1959).

Houcke, Judith. *Hot and Bothered: Women, Medicine, and Menopause in Modern America*. Cambridge: Harvard University Press, 2006.

Howdeshell, Kembra L., Andrew K. Hotchkiss, Kristina A. Thayer, John G. Vandenbergh, and Frederick S. vom Saal. "Exposure to Bisphenol A Advances Puberty." *Nature* 401 (1999): 763–764.

Howdeshell, Kembra L., Paul H. Peterman, Barbara M. Judy, Julia A. Taylor, Carl E. Orazio, Rachel L. Ruhlen, Frederick S. vom Saal, and Wade V. Welshons. "Bisphenol A Is Released from Used Polycarbonate Animal Cages into Water at Room Temperature." *Environmental Health Perspectives* 111 (2003): 1180–1187.

Howe, S. R., L. Borodinsky, and R. S. Lyon. "Potential Exposure to Bisphenol A From Food-Contact Uses of Epoxy Can Coatings." *Journal of Coatings Technology* 70 (1997): 69–74.

Howell, J. M., and C. M. Pickering. "Suspected Synthetic Oestrogen Poisoning in Mink." *Veterinary Record* 76 (1964):169–170.

Huff, James. Review of Devra Davis, *The Secret History of the War on Cancer. Environmental Health Perspectives* 116 (2008): A90.

Hunt, Patricia A., Kara E. Koehler, Martha Susiarjo, Craig A. Hodges, Arlene Ilagan, Robert C. Voigt, Sally Thomas, Brian F. Thomas, and Terry J. Hassold. "Bisphenol A Exposure Causes Meiotic Aneuploidy in the Female Mouse." *Current Biology* 13 (2003): 546–553.

Hurwitz, David. "Pregnancy Accidents in Diabetes." *Journal of the American Medical Association* 116 (1941): 645.

Hurwitz, David, and Katherine Kuder. "Fetal and Neo-Natal Mortality in Pregnancy Complicated by Diabetes Mellitus." *Journal of the American Medical Association* 124 (1944): 271–275.

Imai, Y. "Comments on 'Estrogenicity of Resin-Based Composites and Sealants Used in Dentistry.'" *Environmental Health Perspectives* 107 (1999): A290.

Ingersoll, Bruce. "U.S. Launches Probe After Switzerland Finds Illegal Hormone in American Beef." *Wall Street Journal*, February 2, 2000.

Institute of Medicine, Committee on Understanding the Biology of Sex and Gender Differences, Theresa M. Wizemann and Mary-Lou Pardue, eds. *Exploring the Biological Contributions to Human Health: Does Sex Matter?* Washington, D.C.: National Academy Press, 2001.

Jacobs, Paul. "U.S., Europe Lock Horns in Beef Hormone Debate." *Los Angeles Times*, April 9, 1999.

Jenkins, R., R. A. Angus, H. McNatt, W. M. Howell, J. A. Kemppainen, M. Kirk, and E. M. Wilson. "Identification of Androstenedione in a River Containing

Paper Mill Effluent." *Environmental Toxicology and Chemistry* 20 (2001): 1325–1331.

Jobling, S., S. Coey, J. G. Whitmore, D. E. Kime, K. J. W. Van Look, and B. G. McAllister. "Wild Intersex Roach (*Rutilus rutilus*) Have Reduced Fertility." *Biological Reproduction* 67 (2002): 515–524.

Jobling, S., M. Nolan, C. R. Tyler, G. Brighty, and J. P. Sumpter. "Widespread Sexual Disruption in Wild Fish." *Environmental Science and Technology* 32 (1998): 2498–2506.

Jobling, S., R. Williams, A. Johnson, A. Taylor, M. Gross-Sorokin, M. Nolan, C. R. Tyler, R. van Aerle, E. Santos, and G. Brighty. "Predicted Exposures to Steroid Estrogens in U.K. Rivers Correlate with Widespread Sexual Disruption in Wild Fish Populations." *Environmental Health Perspectives* 114, suppl. 1 (2006): 32–39.

Joffe, Michael. "Infertility and Environmental Pollutants." *British Medical Bulletin* 68 (2003): 47–70.

———. "Myths About Endocrine Disruption and the Male Reproductive System Should Not Be Propagated." *Human Reproduction,* 17 (2002): 520–521.

Johnson, Christine. "Endocrine Disruptors and the Transgendered." *Trans-health* 4 (2002). http://www.trans-health.com/displayarticle.php?aid=51 (accessed July 26, 2008).

Jordan, Andrew, and Timothy O'Riodan. "The Precautionary Principle in Contemporary Environmental Policy and Politics." In *Protecting Public Health and the Environment: Implementing the Precautionary Principle,* ed. Carolyn Raffensperger and Joel Tickner, 15–36. Washington, D.C.: Island, 1999.

Josephson, Julian. "Breaching the Placenta." *Environmental Health Perspectives* 108 (2000): A468.

Jukes, Tom. "Estrogens in Beef Production." *BioScience* 26 (1976): 544–547.

Kabat, Geoffrey. *Hyping Health Risks: Environmental Hazards in Daily Life and the Science of Epidemiology.* New York: Columbia University Press, 2008.

Kaplan, N. M. "Male Pseudohermaphroditism: Report of a Case, with Observations on Pathogenesis." *New England Journal of Medicine* 261 (September 24, 1959): 642–643.

Kaplowitz, Paul B., Eric J. Slora, Richard C. Wasserman, Steven E. Pedlow, and Marcia E. Herman-Giddens. "Earlier Onset of Puberty in Girls: Relation to Increased Body Mass Index and Race." *Pediatrics* 108 (2001): 347–353.

Karnaky, Karl John. "Estrogenic Tolerance in Pregnant Women." *American Journal of Obstetrics and Gynecology* 51 (1947): 312–316.

Kelsey, Frances O. "Historical Perspective." Talk given as part of the workshop *Thalidomide: Potential Benefits and Risks* at the National Institutes of Health, September 9, 1997. http://www.fda.gov/oashi/patrep/nih99.html (accessed July 22, 2008).

Kerlin, Scott P. "The Presence of Gender Dysphoria, Transsexualism, and Disorders of Sexual Differentiation in Males Prenatally Exposed to Diethylstilbestrol: Initial Evidence from a 5-Year Study." Paper presented at 6th Annual

E-Hormone Conference, New Orleans, October 27–30, 2004. http://www.antijen.org/transadvocate/id33.html (accessed July 26, 2008).

Kiesecker, Joseph M. "Synergism Between Trematode Infection and Pesticide Exposure: A Link to Amphibian Limb Deformities in Nature?" *Proceedings of the National Academy of Sciences* 99 (2002): 9900–9904.

King, M. C., J. H. Marks, and J. B. Mandell. "Breast and Ovarian Cancer Risks Due to Inherited Mutations in BRCA1 and BRCA2." *Science* 302 (2003): 643–646.

Kleiner, Kurt. "US May Relax Rules on Carcinogens in Food." *New Scientist* (July 27, 1996). http://www.newscientist.com/article/mg15120401.400-us-may-relax-rules-on-carcinogens-in-food.html (accessed July 23, 2008).

Kolpin, Dana W., Edward T. Furlong, Michael T. Meyer, E. Michael Thurman, Steven D. Zaugg, Larry B. Barber, and Herbert T. Buxton. "Pharmaceuticals, Hormones, and Other Organic Wastewater Contaminants in U.S. Streams, 1999–2000: A National Reconnaissance." *Environmental Science and Technology* 36 (2002): 1202–1211.

Krimsky, Sheldon. "An Epistemological Inquiry into the Endocrine Disruptor Thesis." *Annals of the New York Academy of Sciences* 948 (2001): 130–141.

———. *Hormonal Chaos: The Scientific and Social Origins of the Environmental Endocrine Hypothesis.* Baltimore: Johns Hopkins University Press, 2000.

Kroll-Smith, Steve, and Worth Lancaster. "Bodies, Environments, and a New Style of Reasoning." *Annals of the American Academy of Political and Social Science* 584 (2002): 203–212.

Laitman Orenberg, Cynthia. *DES: The Complete Story.* New York: St. Martin's, 1981.

Lang, Iain A., Tamara S. Galloway, Alan Scarlett, William E. Henley, Michael Depledge, Robert B. Wallace, and David Melzer. "Association of Urinary Bisphenol A Concentration with Medical Disorders and Laboratory Abnormalities in Adults." *Journal of the American Medical Association* 2008 (300): 1303–1310.

Langston, N., S. Freeman, D. Gori, and S. Rohwer. "Evolution of Body Size in Female Redwinged Blackbirds: Effects of Female Competition and Reproductive Energetics." *Evolution* 44 (1990): 1764–1779.

Langston, N., and N. Hillgarth. "The Extent of Primary Molt Varies with Parasites in Laysan Albatrosses: A Possible Role in Life History Tradeoffs Between Current and Future Reproduction." *Proceedings of the Royal Society of London, Series B: Biological Sciences* 261 (1995): 239–243.

Langston, N., and S. Rohwer. "Molt/Breeding Tradeoffs in Albatrosses: Implications for Understanding Life History Variables." *Oikos* 76 (1995): 498–510.

Langston, N., S. Rohwer, and D. Gori. "Experimental Analysis of Intra- and Intersexual Competition in Red-Winged Blackbirds."*Behavioral Ecology* 8 (1997): 524–533.

Langston, Nancy. "Gender Transformed: Endocrine Disruptors in the Environment." In *Seeing Nature Through Gender,* ed. Virginia Scharff, 129–166. Lawrence: University of Kansas Press, 2003.

———. "The Retreat from Precaution: Regulating Diethylstilbestrol (DES), Endocrine Disruptors, and Environmental Health." *Environmental History* 13 (2008): 41–65.

Lapa, Bernard. "Diethylstilbestrol in the Treatment of Idiopathic Repeated Abortion." *New York State Journal of Medicine* 48 (December 1, 1948): 2614.

Latour, Bruno. *We Have Never Been Modern.* Cambridge: Harvard University Press, 1993.

Lear, Linda. *Rachel Carson: Witness for Nature.* New York: Holt, 1998.

Lebel, J. *Health: An Ecosystem Approach.* Ottawa: IDRC Books, 2004.

LeBlanc, Gerald A. "Are Environmental Sentinels Signaling?" *Environmental Health Perspectives* 103 (1995): 888–890.

———. "Steroid Hormone-Regulated Processes in Invertebrates and Their Susceptibility to Environmental Endocrine Disruption." In *Environmental Endocrine Disrupters: An Evolutionary Perspective,* ed. Louis J. Guillette, Jr., and D. Andrew Crain, 126–154. London: Taylor and Francis, 2000.

Lee, Joyce M., and Joel D. Howell. "Tall Girls: The Social Shaping of a Medical Therapy." *Archives of Pediatric and Adolescent Medicine* 160 (2006): 1035–1039.

Leech, Paul Nicholas. "Stilbestrol: Preliminary Report of the Council on Pharmacy and Chemistry (AMA)." *Journal of the American Medical Association* 113 (1939): 2312–2320.

Leonard, Andrew. "How the World Works: Does Plastic Make Us Fat?" *Salon* (July 16, 2007). http://www.salon.com/tech/htww/2007/07/16/obesity/ (accessed July 26 2008).

"Less Paper." *Time* (January 11, 1943). http://www.time.com/time/magazine/article/0,9171,790665,00.html (accessed June 6, 2009).

Lever, J., D. A. Frederick, K. Laird, L. Sadeghi-Azar. "Tall Women's Satisfaction with Their Height: General Population Data Challenge Assumptions Behind Medical Interventions to Stunt Girls' Growth." *Journal of Adolescent Health*40 (2007): 192–194.

Longnecker, L., M. Klebanoff, H. Zhou, and J. Brock. "Association Between Maternal Serum Concentration of the Ddt Metabolite DDE and Preterm and Small-For-Gestational-Age Babies at Birth." *Lancet* 358 (2001): 110–114.

Louhiala, P. "How Tall Is Too Tall? On the Ethics of Oestrogen Treatment for Tall Girls." *Journal of Medical Ethics* 33 (2007): 48–50.

Lowe, Derek. "Testosterone, Carbon Isotopes, and Floyd Landis." *Corante* (August 1, 2006). http://pipeline.corante.com/archives/2006/08/01/testoster one_carbon_isotopes_and_floyd_landis.php (accessed May 31, 2009).

Ma, Risheng, and David A. Sassoon. "PCBs Exert an Estrogenic Effect Through Repression of the Wnt7a Signaling Pathway in the Female Reproductive Tract." *Environmental Health Perspectives* 114 (2006): 898–904.

Marcus, Alan. *Cancer from Beef: DES, Federal Food Regulation, and Consumer Confidence.* Baltimore: Johns Hopkins University Press, 1994.

———. "The Newest Knowledge of Nutrition: Wise Burroughs, DES, and Modern Meat." *Agricultural History* 67 (1993): 66–85.

Markey, C.M., E. H. Luque, M. Muñoz de Toro, C. Sonnenschein, and A. M. Soto. "In Utero Exposure to Bisphenol A Alters the Development and Tissue Organization of the Mouse Mammary Gland." *Biology of Reproduction* 65 (2001): 1215–1223.

Markowitz, G., and D. Rosner. *Deceit and Denial: The Deadly Politics of Industrial Pollution*. Berkeley: University of California Press, 2002.

———. "Industry Challenges to the Principle of Prevention in Public Health: The Precautionary Principle in Historical Perspective." *Public Health Report* 117 (2002): 501–512.

Marks, Harry M. *The Progress of Experiment: Science and Therapeutic Reform in the United States, 1900–1990*. Cambridge: Cambridge University Press, 1997.

Marler, P., S. Peters, G. F. Ball, A. M. Dufty, Jr., and J. C. Wingfield. "The Role of Sex Steroids in the Acquisition and Production of Bird Song." *Nature* 336 (1988): 770–772.

"Masculinity at Risk," *Nature* 375 (1995): 522.

McCoy, Michael, Marc Reisch, and Alexander Tullo. "Facts and Figures of the Chemical Industry." *Chemical and Engineering News* 84 (29): 35–72.

McKiernan, J. M., T. W. Hensle, and H. Fisch. "Increasing Risk of Developing Testicular Cancer by Birth Cohort in the United States." *Dialogues in Pediatric Urology* 23 (2000): 7–8.

McLachlan, J. A., R. R. Newbold, M. E. Burow, and S. F. Li. "From Malformations to Molecular Mechanisms in the Male: Three Decades of Research on Endocrine Disrupters." *APMIS* 109 (2001): 263–272.

McLachlan, John A. "Environmental Signaling: What Embryos and Evolution Teach Us About Endocrine Disrupting Chemicals." *Endocrine Reviews* 22 (2001): 319–341.

McLachlan, John A., ed. *Estrogens in the Environment II: Influences on Development*. New York: Elsevier Science, 1985.

Meharg, Andy. "Science in Culture: The Arsenic Green." *Nature* 423 (2003): 688.

Meikle, Jeffery L. *American Plastic: A Cultural History*. New Brunswick, N.J.: Rutgers University Press, 1995.

Meis, Paul J., Mark Klebanoff, Elizabeth Thom, Mitchell P. Dombrowski, Baha Sibai, Atef H. Moawad, Catherine Y. Spong, John C. Hauth, Menachem Miodovnik, Michael W. Varner, Kenneth J. Leveno, Steve N. Caritis, Jay D. Iams, Ronald J. Wapner, Deborah Conway, Mary J. O'Sullivan, Marshall Carpenter, Brian Mercer, Susan M. Ramin, John M. Thorp, and Alan M. Peaceman. "Prevention of Recurrent Preterm Delivery by 17 Alpha-Hydroxyprogesterone Caproate." *New England Journal of Medicine* 348 (2003): 2379–2385.

Mendola, Pauline, Germaine M. Buck, John E. Vena, Maria Zielezny, and Lowell E. Sever. "Consumption of PCB-Contaminated Sport Fish and Risk of Spontaneous Fetal Death." *Environmental Health Perspectives* 103 (1995): 498–502.

Mergler, D. "Integrating Human Health into an Ecosystem Approach to Mining." In *Managing for Healthy Ecosystems,* ed. D. Rapport, 875–883. Boca Raton, Fla.: CRC Press, Lewis Publishers, 2003.

Mericskay, Mathias, Luca Carta, and David Sassoon. "Diethylstilbestrol Exposure in Utero: A Paradigm for Mechanisms Leading to Adult Disease." *Birth Defects Research A: Clinical and Molecular Teratology* 73 (2005): 133–135.

Metcalfe, Chris D. "Sex And Sewage: Estrogenic Compounds in the Environment." Abstract. Geological Society of American Annual Meeting 2001. http://gsa.confex.com/gsa/2001AM/finalprogram/abstract_25760.htm (accessed June 2, 2009).

Meyerowitz, Joanne. *How Sex Changed: A History of Transsexuality in the United States.* Cambridge: Harvard University Press, 2002.

Meyers, Robert. *DES: The Bitter Pill.* New York: Seaview/Putnum, 1983.

Michaels, David. "Doubt Is Their Product." *Scientific American* 292 (2005): 96–101.

———. "Scientific Evidence and Public Policy." *American Journal of Public Health* 95 (2005): S5–S7.

Michaels, David, and Celeste Monforton. "Manufacturing Uncertainty: Contested Science and the Protection of the Public's Health and Environment." *American Journal of Public Health* 95 (2005): S39–S48.

Miller, C., K. Degenhardt, and D. A. Sassoon. "Fetal Exposure to DES Results in De-Regulation of *Wnt7a* During Uterine Morphogenesis." *Nature Genetics* 20 (1998): 228–230.

Miller, C., and D. A. Sassoon. "*Wnt7a* Maintains Appropriate Uterine Patterning During the Development of the Mouse Female Reproductive Tract." *Development* 125 (1998): 3201-3211.

Milloy, Steven. "Junk Science: Anatomy of a Chemical Murder." FoxNews.com (April 24, 2008). http://www.foxnews.com/story/0,2933,352478,00.html (accessed June 18, 2008).

———. "Junk Science Report: California's Bogus Baby Bottle Scare." *Canada Free Press,* April 25, 2005. http://www.canadafreepress.com/2005/milloy042505.htm (accessed May 30, 2009).

Mitman, Gregg, Michelle Murphy, and Christopher Sellers, eds. "Landscapes of Exposure: Knowledge and Illness in Modern Environments." *Osiris* 19 (2004).

Mittwoch, Ursula. "Sex Determination in Mythology and History." *Arquivos Brasileiros de Endocrinologia & Metabologia* 49 (2005). doi: 10.1590/S0004-27302005000100003.

Miyakoda, H., M. Tabata, S. Onodera, and K. Takeda. "Passage of Bisphenol A into the Fetus of the Pregnant Rat." *Journal of Health Science* 45 (1999): 318–323.

Moore, Patrick. "Why I Left Greenpeace." *Wall Street Journal Opinion Journal,* April 22, 2008. http://online.wsj.com/article/SB120882720657033391.html?mod=opinion_main_commentaries (accessed July 27, 2008).

Myers, John Peterson. "Good Genes Gone Bad." *American Prospect* (March 19, 2006). http://www.prospect.org/cs/articles?article=good_genes_gone_bad (accessed July 21, 2008).

———. "Hormonally Active Agents in the Environment." Our Stolen Future (1999). http://ourstolenfuture.org/Consensus/nrc.htm (accessed July 21, 2008).

———. "Why Endocrine Disruption Challenges Current Approaches to Regulation of Chemicals." Our Stolen Future. http://www.ourstolenfuture.org/Basics/challenge.htm (accessed Jan. 15, 2008).

Mylchreest, Eva, Duncan G. Wallace, Russell C. Cattley, and Paul M. D. Foster. "Dose-Dependent Alterations in Androgen-Regulated Male Reproductive Development in Rats Exposed to Di(n-butyl) Phthalate During Late Gestation." *Toxicological Sciences* 55 (2000): 143–151.

Nagel, S. C., F. S. vom Saal, K. A. Thayer, M. G. Dhar, M. Boechler, and W. V. Welshons. "Relative Binding Affinity-Serum Modified Access (RBA-SMA) Assay Predicts the Relative in Vivo Bioactivity of the Xenoestrogens Bisphenol A and Octylphenol." *Environmental Health Perspectives* 105 (1997): 70–76.

Nagler, James J., Jerry Bouma, Gary H. Thorgaard, and Dennis D. Dauble. "High Incidence of a Male-Specific Genetic Marker in Phenotypic Female Chinook Salmon from the Columbia River." *Environmental Health Perspectives* 109 (2001): 67–69.

Nash, Linda. *Inescapable Ecologies: A History of Environment, Disease, and Knowledge.* Berkeley: University of California Press, 2007.

———. "Purity and Danger: Historical Reflections on the Regulation of Environmental Pollutants." *Environmental History* 13 (2008): 651–658.

National Toxicology Program, U.S. Department of Health and Human Services. "Endocrine Disruptors Low-Dose Peer Review Final Report." http://ntp.niehs.nih.gov/ntp/htdocs/liason/LowDosePeerFinalRpt.pdf (accessed June 6, 2009).

———. "NTP-CERHR Expert Panel Report on the Reproductive and Developmental Toxicity of Bisphenol A." November 26, 2007. http://cerhr.niehs.nih.gov/chemicals/bisphenol/BPAFinalEPVF112607.pdf (accessed July 26, 2008).

Needleman, H. L. "The Case of Deborah Rice: Who Is the Environmental Protection Agency Protecting?" *PLoS Biology* 6 (2008): e129. doi:10.1371/journal.pbio.0060129

Neustadt, Robert E., and Ernest R. May. *Thinking in Time: The Uses of History for Decision Makers.* New York: Free Press, 1986.

Newbold, R. R. "Gender-Related Behavior in Women Exposed Prenatally to Diethylstilbestrol." *Environmental Health Perspectives* 101 (1993): 208–213.

———. "Lessons Learned from Perinatal Exposure to Diethylstilbestrol." *Toxicology and Applied Pharmacology* 199 (2004): 142–150.

———. "Perinatal Carcinogenesis: Growing a Node for Epidemiology, Risk Management, and Animal Studies." *Toxicology and Applied Pharmacology* 199 (2004): 142–150.

Newbold, R. R., W. N. Jefferson, and E. Padilla-Banks. "Long-Term Adverse Effects of Neonatal Exposure to Bisphenol A on the Murine Female Reproductive Tract." *Reproductive Toxicology* 24 (2007): 253–258.

Newbold, R. R., E. Padilla-Banks, and W. N. Jefferson. "Adverse Effects of the Model Environmental Estrogen Diethylstilbestrol Are Transmitted to Subsequent Generations." *Endocrinology* 147 (2006): s11–s17.

Newbold, R. R., E. Padilla-Banks, R. J. Snyder, and W. N. Jefferson. "Developmental Exposure to Estrogenic Compounds and Obesity." *Birth Defects Research A: Clinical and Molecular Teratology* 73 (2005): 478–480.

Newbold, R. R., E. Padilla-Banks, R. J. Snyder, T. M. Phillips, and W. N. Jefferson. "Developmental Exposure to Endocrine Disruptors and the Obesity Epidemic." *Reproductive Toxicology* 23 (2007): 290–296.

Nilsson, E. E., M. D. Anway, J. Stanfield, and M. K. Skinner. "Transgenerational Epigenetic Effects of the Endocrine Disruptor Vinclozolin on Pregnancies and Female Adult Onset Disease." *Reproduction* 135 (2008): 713–721.

Noble, R. L. "Functional Impairment of the Anterior Pituitary Gland Produced by the Synthetic Oestrogenic Substance 4:4′ dihydroxy-á:â-diethylstilbene." *Journal of Physiology* 94 (1938): 177–183.

Nurminen, T. "The Epidemiologic Study of Birth Defects and Pesticides." *Epidemiology* 12 (2001): 145–146.

Nyhart, Lynn K. *Biology Takes Form: Animal Morphology and the German Universities, 1800–1900*. Chicago: University of Chicago Press, 1995.

O'Brien, Mary. *Making Better Environmental Decisions: An Alternative to Risk Assessment*. Cambridge: MIT Press, 2000.

Olea, N. "Comments on 'Estrogenicity of Resin-Based Composites and Sealants Used in Dentistry': Response." *Environmental Health Perspectives* 107 (1999): A290–A292.

Olea, N., R. Pulgar, P. Perez, F. Olea-Serrano, A. Rivas, A. Novillo-Fertrell, V. Pedraza, A. M. Soto, and C. Sonnenschein "Estrogenicity of Resin-Based Composites and Sealants Used in Dentistry." *Environmental Health Perspectives* 104 (1996): 298–305.

Oreskes, Naomi. "The American Denial of Global Warming," University of California Television, December 12, 2007/ http://www.uctv.tv/search-details.asp?showID=13459 (accessed Nov. 29, 2008).

Orlando, E. F., A. S. Kolok, G. A. Binzcik, J. L. Gates, M. K. Horton, C. S. Lambright, L. E. Gray, Jr., A. M. Soto, and L. J. Guillette, Jr. "Endocrine-Disrupting Effects of Cattle Feedlot Effluent on an Aquatic Sentinel Species, the Fathead Minnow." *Environmental Health Perspectives* 112 (2004): 353–358.

Oudshoorn, Nelly. *Beyond the Natural Body: An Archaeology of Sex Hormones*. London: Routledge, 1994.

Palmer J. R., L. A. Wise, E. E. Hatch, R. Troisi, L. Titus-Ernstoff, W. Strohsnitter, R. Kaufman, A. L. Herbst, K. L. Noller, H. Hyer, and R. N. Hoover. "Prenatal Diethylstilbestrol Exposure and Risk of Breast Cancer." *Cancer Epidemiology Biomarkers and Prevention* 15 (2006): 1509–1514.

Palmlund, I. "Exposure to a Xenoestrogen Before Birth: The Diethylstilbestrol Experience." *Journal of Psychosomatic Obstetrics and Gynecology* 17 (1996): 71–84.

Parkes, A. S., E. C. Dodds, and R. L. Noble. "Interruption of Early Pregnancy by Means of Orally Active Estrogens." *British Medical Journal* 2 (1938): 557–559.

Paulozzi, L. J. "International Trends in Rates of Hypospadias and Cryptorchidism." *Environmental Health Perspectives* 107 (1999): 297–302.

Paulozzi, L. J., J. D. Erickson, and R. J. Jackson. "Hypospadias Trends in Two U.S. Surveillance Systems." *Pediatrics* 100 (1997): 831–834.

Paulsen, C. A., N. G. Berman, and C. Wang. "Data from Men in Greater Seattle Reveals No Downward Trend in Semen Quality: Further Evidence that Deterioration of Semen Quality Is Not Geographically Uniform." *Fertility and Sterility* 65 (1996): 1015–1020

Pearce, N. "Corporate Influences on Epidemiology." *International Journal of Epidemiology* 37 (2008): 46–53.

——. "Response: The Distribution and Determinants of Epidemiologic Research." *International Journal of Epidemiology* 37 (2008): 65–68.

Perez Comas, A. "Precocious Sexual Development in Puerto Rico." *Lancet* 8284 (1982): 1299–1300.

Persky, V., M. Turyk, H. A. Anderson, L. P. Hanrahan, C. Falk, D. N. Steenport, R. Chatterton, Jr., and S. Freels. "The Effects of PCB Exposure and Fish Consumption on Endogenous Hormones." *Environmental Health Perspectives* 109 (2001): 1275–1283.

Pollan, Michael. *The Omnivore's Dilemma: A Natural History of Four Meals.* New York: Penguin, 2006.

"Prevent the Occurrence of Toxins in Water to Protect Food and the Environment: 2002 Annual Report." Agricultural Research Service. http://www.ars.usda.gov/research/projects/projects.htm?ACCN_NO=404372&fy=2002 (accessed July 26, 2008).

Proctor, Robert N. *Cancer Wars: How Politics Shapes What We Know and Don't Know About Cancer.* New York: Basic, 1995.

"Progesterone Shots May Prevent Preterm Births." *Family Practice News* 33 (2003): 38–29.

Program on Breast Cancer and Environmental Risk Factors, Sprecher Institute for Comparative Cancer Research, Cornell University. "Consumer Concerns About Hormones in Food." Fact Sheet No. 37 (June 2000). http://envirocancer.cornell.edu/FactSheet/Diet/fs37.hormones.cfm (accessed July 26, 2008).

Purdom, C. E., P. A. Hardiman, V. J. Bye, N. C. Eno, C. R. Tyler, and J. P. Sumpter.

"Estrogenic Effects of Effluents from Sewage-Treatment Works." *Chemistry and Ecology* 8 (1994): 275–285

Rackl, Lorilyn. "A Shot of Hope: Progesterone Injections Are Making a Comeback to Prevent Premature Births." (Arlighton Heights, Ill.) *Daily Herald,* October 27, 2003.

Raffensperger, Carolyn, and Joel Tickner, eds., *Protecting Public Health and the Environment: Implementing the Precautionary Principle.* Washington, D.C.: Island, 1999.

Raloff, Janet. "Common Pollutants Undermine Masculinity." *Science News* 155 (April 3, 1999): 213.

———. "The Gender Benders: Are Environmental 'Hormones' Emasculating Wildlife?" *Science News* 145 (January 8, 1994): 24.

———. "Hormones: Here's the Beef: Environmental Concerns Reemerge over Steroids Given to Livestock." *Science News* 162 (January 5, 2002): 10.

———. "Macho Waters: Some River Pollution Spawns Body-Altering Steroids." *Science News* 159 (January 6, 2001): 8–10.

———. "That Feminine Touch: Are Men Suffering from Prenatal Exposure or Childhood Exposure to 'Hormonal' Toxicants?" *Science News* 145 (1994): 56–59.

Raun, A. P., and R. L. Preston. "History of Diethylstilbestrol Use in Cattle." *American Society of Animal Science* (2002). http://www.asas.org/Bios/Raun hist.pdf (accessed July 23, 2008).

Rea, W. J. *Chemical Sensitivity: Principles and Mechanisms.* Vol. 3 : *Clinical Manifestations of Pollutant Overload.* Boca Raton, Fla.: CRC Press, Lewis Publishers, 1996.

Richtera, C. A., L. S. Birnbaumb, F. Farabollinic, R. R. Newbold, B. S. Rubine, C. E. Talsnessf, J. G. Vandenbergh, D. R. Walser-Kuntz, and F. S. vom Saal. "In Vivo Effects of Bisphenol A in Laboratory Rodent Studies." *Reproductive Toxicology* 24 (2007): 199–224.

Roberts, Celia. "Biological Behavior? Hormones, Psychology, and Sex." *NWSA Journal* 12 (2000): 1–20.

———. "Drowning in a Sea of Estrogens: Sex Hormones, Sexual Reproduction, and Sex." *Sexualities* 6 (2003): 195–213.

———. "'A Matter of Embodied Fact': Sex Hormones and the History of Bodies." *Feminist Theory* 3 (2001): 7–26.

Roberts, David. "Uncovering the Weinberg Group," *Vanity Fair,* April 28, 2008, http://www.vanityfair.com/online/daily/2008/04/uncovering-the.html (accessed June 2, 2009).

Roberts, Jody A., and Nancy Langston. "Toxic Bodies/Toxic Environments: An Interdisciplinary Forum." *Environmental History* 13 (2008): 629–635.

Rodgers-Gray, T. P., S. Jobling, S. Morris, C. Kelly, S. Kirby, A. Janbakhsh, J. E. Harries, M. J. Waldock, J. P. Sumpter, and C. R. Tyler. "Long-Term Temporal Changes in the Estrogenic Composition of Treated Sewage Effluent and Its

Biological Effects on Fish." *Environmental Science and Technology* 34 (2000): 1521–1528.

Rohwer, S., N. Langston, and D. Gori. "Body Size in Male Redwinged Blackbirds: Manipulating Selection with Sex-Specific Feeders." *Evolution* 50 (1996): 2049–2065.

Rosario, Vernon. "The Biology of Gender and the Construction of Sex?" *GLQ: A Journal of Lesbian and Gay Studies* 10 (2004): 280–287.

Rosenberg, J. "Low Birth Weight Is Linked to Timing of Prenatal Care and Other Maternal Factors." *International Family Planning Perspectives* 30, no. 2 (June 2004). http://www.guttmacher.org/pubs/journals/3010104.html (accessed July 27, 2008).

Rosenblum, G., and E. Melinkoff. "Preservation of the Threatened Pregnancy with Particular Reference to the Use of Diethylstilbestrol." *Western Journal of Surgery* (1947): 601–603.

Roughgarden, Joan. *Evolution's Rainbow: Diversity, Gender, and Sexuality in Nature and People.* Berkeley: University of California Press, 2004.

Rubin, B. S., J. R. Lenkowski, C. M. Schaeberle, L. N. Vandenberg, P. M. Ronsheim, and A. M. Soto. "Evidence of Altered Brain Sexual Differentiation in Mice Exposed Perinatally to Low, Environmentally Relevant Levels of Bisphenol A." *Endocrinology* 147 (2006): 3681–3691.

Rudacille, Deborah. *The Riddle of Gender: Science, Activism, and Transgender Rights.* New York: Pantheon, 2005.

Russell, Edmund. *War and Nature: Fighting Humans and Insects with Chemicals from World War I to "Silent Spring."* Cambridge: Cambridge University Press, 2001.

Rust, Susanne, Meg Kissinger, and Cary Spivak. "Are Your Products Safe? You Can't Tell." *Milwaukee Journal Sentinel,* November 24, 2007. http://www.jsonline.com/watchdog/watchdogreports/29331224.html (accessed June 7, 2009).

Safe, Stephen. "Environmental and Dietary Estrogens and Human Health: Is There a Problem? *Environmental Health Perspectives* 103 (1995): 346–351.

Saldeen, Pia, and Tom Saldeen. "Omega-3 Fatty Acids: Structure, Function, and Relation to the Metabolic Syndrome, Infertility, and Pregnancy." *Metabolic Syndrome and Related Disorders* 4 (2006): 138–148.

Sar, M., E. Mylchreest, and P. M. D. Foster. "Di(n-butyl) Phthalate Induces Changes in Morphology and Androgen Receptor Levels in the Fetal Testis." Paper presented at the Meeting of the Society of Toxicology, New Orleans, March 1999.

Schell, Orville. *Modern Meat: Antibiotics, Hormones, and the Pharmaceutical Farm.* New York: Random House, 1984.

Schettler, Ted. "Endocrine Disruptors: The State of the Science." Greater Boston Physicians for Social Responsibility (1997). http://web.archive.org/web/19990220214027/www.psr.org/tedfs.htm (accessed July 18, 2008).

Scott, James C. *Seeing Like a State: How Certain Schemes to Improve the Human Condition Have Failed.* New Haven: Yale University Press, 1998.

Seaman, Barbara. *The Greatest Experiment Ever Performed on Women: Exploding the Estrogen Myth.* New York: Hyperion, 2003.

Seidman, Lisa A., and Noreen Warren. "Frances Kelsey and Thalidomide in the US: A Case Study Relating to Pharmaceutical Regulations." *American Biology Teacher* 64 (2002): 495–500.

Sellers, Christopher. *Hazards of the Job: From Industrial Disease to Environmental Health Science.* Chapel Hill: University of North Carolina Press, 1997.

———. "Thoreau's Body: Towards an Embodied Environmental History." *Environmental History* 4 (1999): 486–514.

Sharpe, Richard, and Niels Skakkebaek. "Are Oestrogens Involved in Falling Sperm Counts and Disorders of the Male Reproductive Tract?" *Lancet* 341 (1993): 1392–1395.

Sheehan, D. M. "No-Threshold Dose-Response Curves for Nongenotoxic Chemicals: Findings and Applications for Risk Assessment." *Environmental Research* 100 (2006): 93–99.

Shubik, Philippe. "Potential Carcinogenicity of Food Additives and Contaminants." *Cancer Research* 35 (1975): 3475–3480.

Silbergeld, E. K. "Risk Assessment: The Perspective and Experience of U.S. Environmentalists." *Environmental Health Perspectives* 101 (1993): 100–104.

Silverman, W. "The Schizophrenic Career of a 'Monster Drug.'" *Pediatrics* 110 (2002): 404–406.

Singer, Natasha. "Does Testosterone Build a Better Athlete?" *New York Times,* August 10, 2006.

Singleton, D. W., Y. Feng, J. Yang, A. Puga, A. V. Lee, and S. A. Khan. "Gene Expression Profiling Reveals Novel Regulation by Bisphenol-A in Estrogen Receptor–Positive Human Cells." *Environmental Research* 100 (2006): 86–92.

Smith, Olive. "Diethylstilbestrol in the Prevention and Treatment of Complications of Pregnancy." *American Journal of Obstetrics and Gynecology* 56 (1948): 821–834.

Smith, Olive, and George Smith. "Prolan and Estrin in the Serum and Urine of Diabetic and Nondiabetic Women During Pregnancy, with Especial Reference to Late Pregnancy Toxemia." *American Journal of Obstetrics and Gynecology* 33 (1937): 365–379.

Smith, Olive, George Smith, and David Hurwitz, "Increased Excretion of Pregnanediol in Pregnancy with Diethylstilbestrol with Special Reference to the Prevention of Late Pregnancy Accidents." *American Journal of Obstetrics and Gynecology* 51 (1946): 411–415.

Soto, A. M., H. Justicia, J. W. Wray, and C. Sonnenschein. "P-Nonyl-phenol: An Estrogenic Xenobiotic Released from 'Modified' Polystyrene." *Environmental Health Perspectives* 92 (1991): 167–173.

Soto, Ana. Autobiographical essay. *in-cites* (October 2001). http://www.in-cites.com/papers/dr-ana-soto.html.

Spearow, J. L., P. Doemeny, R. Sera, R. Leffler, M. Barkley. "Genetic Variation in Susceptibility to Endocrine Disruption by Estrogen in Mice." *Science* 285 (1999): 1259–1261.

Steingraber, Sandra. *Having Faith: An Ecologist's Journey to Motherhood*. New York: Perseus, 2001.

———. *Living Downstream: An Ecologist Looks at Cancer and the Environment*. New York: Da Capo, 1997.

Stephens, T., and R. Brynner. *Dark Remedy: The Impact of Thalidomide and Its Revival as a Vital Medicine*. Cambridge, Mass.: Perseus, 2001.

Sugiura-Ogasawara, M., Y. Ozaki, S. Sonta, T. Makino, and K. Suzumori. "Exposure to Bisphenol A Is Associated with Recurrent Miscarriage." *Human Reproduction Online* (June 9, 2005). http://humrep.oxfordjournals.org/cgi/content/abstract/deh888v1 (accessed July 21, 2008).

Summons, T. G. "Animal Feed Additives, 1940–1966." *Agricultural History* 42 (1968): 305–313.

Sundqvist C., A. G. Amador, and A. Bartke. "Reproduction and Fertility in the Mink (*Mustela vision*)." *Journal of Reproduction and Fertility* 85 (1989): 413–441.

Susiarjo, M., T. J. Hasold, E. Freeman, and P. A. Hunt. "Bisphenol A Exposure in Utero Disrupts Early Oogenesis in the Mouse." *PLoS Genetics* 3 (2007): e5.

Suzuki, A., H. Urushitani, T. Sato, T. Kobayashi, H. Watanabe, Y. Ohta, and T. Iguchi. "Gene Expression Change in the Müllerian Duct of the Mouse Fetus Exposed to Diethylstilbestrol in Utero." *Experimental Biology and Medicine* 232 (2007): 503–514

Swan, S. H., E. P. Elkin, and L. Fenster. "Have Sperm Densities Declined? A Reanalysis of Global Trend Data." *Environmental Health Perspectives* 105 (1997): 1228–1232.

———. "The Question of Declining Sperm Density Revisited: An Analysis of 101 Studies Published 1934–1996." *Environmental Health Perspectives* 108 (2000): 961–966.

Swan, S. H., F. Liu, J. W. Overstreet, C. Brazil, and N. E. Skakkebaek. "Semen Quality of Fertile US Males in Relation to Their Mothers' Beef Consumption During Pregnancy." *Human Reproduction* 22 (2007): 1497–1502.

Swyer, G. I. M., and R. G. Law. "An Evaluation of the Prophylactic Ante-Natal Use of Stilboestrol: Preliminary Report." *Journal of Endocrinology* 10 (1954): vi.

"Synthetic Female Hormone Pills Considered Potential Danger." *Science News Letter* (January 13, 1940).

Sze, Julie. "Boundaries and Border Wars: DES, Technology, and Environmental Justice." *American Studies Quarterly* 58 (2006): 791–814.

Szyf, Moshe. "The Dynamic Epigenome and Its Implications in Toxicology." *Toxicological Sciences* 100 (2007): 7–23.

Takahashi, O., and S. Oishi. "Disposition of Orally Administered 2,2-Bis (4-hydroxyphenyl) Propane (Bisphenol A) in Pregnant Rats and the Placental Transfer to Fetuses." *Environmental Health Perspectives* 108 (2000): 931–935.

Tancrède, C. "The Role of Human Microflora in Health and Disease." *European Journal of Clinical Microbiology and Infectious Diseases* 11 (1992): 1012–1015.

Tanmahasamut, P., J. Liu, L. B. Hendry, and N. Sidell. "Conjugated Linoleic Acid Blocks Estrogen Signaling in Human Breast Cancer Cells." *Journal of Nutrition* 134 (2004): 674–680.

Tannock, G. W. *Normal Microflora: An Introduction to Microbes Inhabiting the Human Body.* Berlin: Springer Berlin, 1994.

Thacker, Paul. "The Weinberg Proposal." News and Features. *Environmental Science and Technology* 40 (2006): 2868–2869.

Thornton, Joe. *Pandora's Poison: Chlorine, Health, and a New Environmental Strategy.* Cambridge: MIT Press, 2000.

Tiefer, Leonore. "Hormone Mistreatment." Review of Anne Fausto-Sterling, *Sexing the Body. Women's Review of Books* 17 (2000): 8–9.

Todd, Julie, and Sorrel Brown. "Pesticide Reform of the Delaney Clause." *Integrated Crop Management News* 24 (October 7, 1996): IC-476. http://www.ipm.iastate.edu/ipm/icm/1996/10-7-1996/pestref.html (accessed June 6, 2009).

Tone, Andrea, and Elizabeth Siegel Watkins, eds. *Medicating Modern America: Prescription Drugs in History.* New York: New York University Press, 2007.

Toppari, J., J. C. Larsen, P. Christiansen, A. Giwercman, P. Grandjean, L. J. Guillette, Jr., B. Jégou, T. K. Jensen, P. Jouannet, N. Keiding, H. Leffers, J. A. McLachlan, O. Meyer, J. Müller, E. Rajpert-De Meyts, T. Scheike, R. Sharpe, J. Sumpter, and N. E. Skakkebaek. *Male Reproductive Health and Environmental Chemicals with Estrogenic Effects.* Copenhagen: Danish Environmental Protection Agency, 1995.

———. "Male Reproductive Health and Environmental Xenoestrogens." *Environmental Health Perspectives* 104, suppl. 4 (1996): 741–806.

Toyama, Y., M. Ohkawa, R. Oku, M. Maekawa, and S. Yuasa. "The Neonatally Administered Diethylstilbestrol Retards the Development of the Blood-Testis Barrier in the Rat." *Journal of Andrology* 22 (2002): 413–423.

Travis, John. "Modus Operandi of an Infamous Drug." *Science News Online* (February 20, 1999). http://www.sciencenews.org/sn_arc99/2_20_99/bob2.htm (accessed July 18, 2008).

Tugwell, Rexford G. *The Diary of Rexford G. Tugwell: The New Deal, 1932–1935,* ed. Michael Vincent Namorato. New York: Greenwood, 1992.

United States Mission to the European Union. "A Primer on Beef Hormones." Statement released by the U.S. Interagency Task Force on Beef Hormones, February 26, 1999. http://useu.usmission.gov/Dossiers/Beef_Hormones/Feb2699_Primer.asp (accessed July 26, 2008).

Venn, A., F. Bruinsma, G. Werther, and P. Pyett. "Oestrogen Treatment to Reduce the Adult Height of Tall Girls: Long-Term Effects on Fertility." *Lancet* 364 (2004): 1513–1519.

Vileisis, Ann. *Kitchen Literacy: How We Lost Knowledge of Where Food Comes from and Why We Need to Get It Back.* Washington, D.C.: Island, 2007.

Visser, Melvin J. *Cold, Clear, and Deadly: Unraveling a Toxic Legacy.* East Lansing: Michigan State University Press, 2007.

Vogel, Sarah. "Battles over Bisphenol A." SKAPP: Case Studies in Science Policy. DefendingScience.org (April 16, 2008). http://www.defendingscience.org/case_studies/Battles-Over-Bisphenol-A.cfm (accessed August 13, 2008).

———. "Chemical Hearing Pits Senators Against FDA." The Pump Handle (May 23, 2008). http://thepumphandle.wordpress.com/2008/05/23/chemical-hearing-pits-senators-against-fda/ (accessed August 12, 2008).

———. "From 'the Dose Makes the Poison' to 'the Timing Makes the Poison': Conceptualizing Risk in the Synthetic Age." *Environmental History* 13 (2008): 667–673.

———. "The Politics of Plastics: The Economic, Political, and Scientific History of Bisphenol A." Ph.D. diss. Columbia University, 2008.

vom Saal, F. S. "Could Hormone Residues Be Involved?" *Human Reproduction* 22 (2007): 1503–1505.

———. "Sexual Differentiation in Litter Bearing Mammals: Influence of Sex of Adjacent Fetuses in Utero." *Journal of Animal Science* 67 (1989): 1824–1840.

vom Saal F. S., B. T. Akingbemi, S. M. Belcher, L. S. Birnbaum, D. A. Crain, E. Eriksen, F. Farabollini, L. J. Guillette, Jr., R. Hauser, J. J. Heindel, S. M. Ho, P. A. Hunt, T. Iguchi, S. Jobling, J. Kanno, R. A. Keri, K. E. Knudsen, H. Laufer, G. A. LeBlanc, M. Marcus, J. A. McLachlan, J. P. Myers, A. Nadal, R. R. Newbold, N. Olea, G. S. Prins, C. A. Richter, B. S. Rubin, C. Sonnenschein, A. M. Soto, C. E. Talsness, J. G. Vandenbergh, L. N. Vandenberg, D. R. Walser-Kuntz, C. S. Watson, W. V. Welshons, Y. Wetherill, and R. T. Zoeller. "Chapel Hill Bisphenol A Expert Panel Consensus Statement: Integration of Mechanisms, Effects in Animals and Potential to Impact Human Health at Current Levels of Exposure." *Reproductive Toxicology* 24 (2007): 131–138.

vom Saal, F. S. and F. Bronson. "Sexual Characteristics of Adult Female Mice Are Correlated with Their Blood Testosterone Levels During Prenatal Development." *Science* 20 (1980): 597–599.

vom Saal, F. S., P. S. Cooke, D. L. Buchanan, P. Palanza, K. A. Thayer, S. Parmigiani, and W. V. Welshons. "A Physiologically Based Approach to the Study of Bisphenol A and Other Estrogenic Chemicals on the Size of Reproductive Organs, Daily Sperm Production, and Behavior." *Toxicology and Industrial Health* 14 (1998): 239–260.

vom Saal, F. S., and C. Hughes. "An Extensive New Literature Concerning Low-Dose Effects of Bisphenol A Shows the Need for a New Risk Assessment." *Environmental Health Perspectives* 113 (2005): 926–933.

vom Saal, F. S., and J. P. Myers. "Editorial: Bisphenol A and Risk of Metabolic Disorders." *Journal of the American Medicial Association* 2008 (300): 1353–1355.

vom Saal, F. S., S. C. Nagel, B. G. Timms, and W. V. Welshons. "Implications for Human Health of the Extensive Bisphenol A Literature Showing Adverse

Effects at Low Doses: A Response to Attempts to Mislead the Public." *Toxicology* 212 (2005): 244–252.
vom Saal, F. S., C. A. Richter, R. R. Ruhlen, S. C. Nagel, B. G. Timms, and W. V. Welshons. "The Importance of Appropriate Controls, Animal Feed, and Animal Models in Interpreting Results From Low-Dose Studies of Bisphenol A." *Birth Defects Research A: Clinical and Molecular Teratology* 73 (2005): 140–145.
vom Saal, F. S., and D. M. Sheehan. "Challenging Risk Assessment." *Forum for Applied Research and Public Policy* 13 (1998): 11–18.
vom Saal, F. S., and W. V. Welshons. "Large Effects from Small Exposures: II. The Importance of Positive Controls in Low-Dose Research on Bisphenol A." *Environmental Research* 100 (2006): 50–76.
Wade N. "DES: A Case Study of Regulatory Abdication." *Science* 177 (1972): 335–337.
Wake, Marvalee H. "Integrative Biology: The Nexus of Development, Ecology, and Evolution." Plenary lecture given at the International Union of Biological Science General Assembly, January 18, 2004, Cairo, Egypt. http://www.iubs.org/test/bioint/46/Wake%20M%20IUBS%20GA%20Plenary-%20Final%20Corected%20MW.htm (accessed July 21, 2008).
Wargo, John. *Our Children's Toxic Legacy: How Science and Law Fail to Protect Us from Pesticides.* New Haven: Yale University Press, 1996.
Watkins, Elizabeth Siegel. *The Estrogen Elixir: A History of Hormone Replacement in America.* Baltimore: Johns Hopkins University Press, 2007.
Waxman, S. H., V. C. Kelley, and S. M. Gartler. "Apparent Masculinization of Female Fetus Diagnosed as True Hermaphrodism by Chromosomal Studies." *Journal of Pediatrics* 60 (1962): 540.
Weaver, Ian C. G. "Epigenetic Programming by Maternal Behavior and Pharmacological Intervention: Nature Versus Nurture; Let's Call the Whole Thing Off." *Epigenetics* 2 (2007): 22–28.
Weinhold, Bob. "Forum: CDC Unveils Body Burden." *Environmental Health Perspectives* 109 (May 2001): A203.
Weisburger, J. H. "Does the Delaney Clause of the U.S. Food and Drug Laws Prevent Human Cancers?" *Toxicological Sciences* 22 (1994): 483–493.
——. "Human Protection Against Non-Genotoxic Carcinogens in the US Without the Delaney Clause." *Experimental and Toxicological Pathology* 48 (1996): 201–208.
——. "The 37 Year History of the Delaney Clause." *Experimental and Toxicological Pathology* 48 (1996): 183–188.
Weiss, Bernard. "Sexually Dimorphic Nonreproductive Behaviors as Indicators of Endocrine Disruption." *Environmental Health Perspectives* 110, suppl. 3 (2002): 387–391.
Welch, R. M., W. Levin, K. Kuntzman, M. N. Jacobson, and A. H. Conney. "Effect of Halogenated Hydrocarbon Insecticides on the Metabolism and Uterotropic Action of Estrogens in Rats and Mice." *Toxicology and Applied Pharmacology* 19 (1971): 234–246.

White, P. "Diabetes Complicating Pregnancy." *Journal of the American Medical Association* 128 (1945): 181–182.

White, P., R. S. Titus, E. P. Joslin, and H. Hunt. "Prediction and Prevention of Late Pregnancy Accidents in Diabetes." *American Journal of Medical Science* 198 (1939): 482–492.

White, Priscilla, and Hazel Hunt. "Pregnancy Complicating Diabetes." *Journal of Clinical Endocrinology* 3 (1943): 500–511.

Whorton, James. *Before "Silent Spring": Pesticides and Public Health in Pre-DDT America.* Princeton: Princeton University Press, 1981.

Williams, G. M, E. Karbe, P. Fenner-Crist, M. J. Iatropoulos, and J. H Weisburger. "Risk Assessment of Carcinogens in Food with Special Consideration of Non-Genotoxic Carcinogens: Scientific Arguments for Use of Risk Assessment and for Changing the Delaney Clause Specifically." *Experimental and Toxicological Pathology* 48 (1996): 209–215.

"Wingspread Statement on the Precautionary Principle." Wingspread Conference, Racine, Wis., January 1998. http://www.gdrc.org/u-gov/precaution-3.html (accessed August 14, 2008).

Woodard, Geoffrey, Ruth R. Ofner, and Charles M. Montgomery. "Accumulation of DDT in the Body Fat and Its Appearance in the Milk of Dogs." *Science* 102 (1945): 177–178.

Woodwell, George. "Toxic Food Web." In *Life Stories: World-Renowned Scientists Reflect on Their Lives and the Future of Life on Earth,* ed. Heather Newbold, 74–84. Berkeley: University of California Press, 2000.

Yarsley, V. E., and E. G. Couzens. "The Expanding Age of Plastics." *Science Digest* 10 (December 1941): 57–59.

Yin, Yan, Congxing Lin, and Liang Ma. "MSX2 Promotes Vaginal Epithelial Differentiation and Wolffian Duct Regression and Dampens the Vaginal Response to Diethylstilbestrol." *Molecular Endocrinology* 20 (2006): 1535–1546.

Yucel, Selcuk, Christopher Dravis, Nilda Garcia, Mark Henkemeyer, and Linda A. Baker. "Hypospadias and Anorectal Malformations Mediated By Eph/Ephrin Signaling." *Journal of Pediatric Urology* 3 (2007): 354–363.

Zebroski, Bob. Review of Andrea Tone and Elizabeth Siegel Watkins, eds., *Medicating Modern America. Journal of the History of Medicine and Allied Sciences* 63 (2008): 4.

Zsarnovszky, A., H. H. Le, H. Wang, and S. M. Belcher. "Ontogeny of Rapid Estrogen-Mediated Extracellular Signal-Regulated Kinase Signaling in the Rat Cerebellar Cortex: Potent Nongenomic Agonist and Endocrine Disrupting Activity of the Xenoestrogen Bisphenol A." *Endocrinology* 146 (2006): 5388–5396.

Zuckerman, Diana. "When Little Girls Become Women: Early Onset of Puberty in Girls," *The Ribbon: A Newsletter of the Cornell University Program on Breast Cancer and Environmental Risk Factors in New York State* 6 (2001): 6–8. http://envirocancer.cornell.edu/Newsletter/pdf/v6i1.pdf (accessed June 6, 2009).

INDEX